The Fats of Life

The Fats of Life

Essential Fatty Acids
in Health and Disease

GLEN D. LAWRENCE

RUTGERS UNIVERSITY PRESS

NEW BRUNSWICK, NEW JERSEY, AND LONDON

LIBRARY OF CONGRESS CATALOGING-IN-PUBLICATION DATA

Lawrence, Glen D., 1948–
    The fats of life : essential fatty acids in health and disease / Glen D. Lawrence.
        p.   cm.
    Includes bibliographical references and index.
    ISBN 978–0-8135–4677–3 (hardcover : alk. paper)
    1. Fatty acids in human nutrition.   2. Essential fatty acids—Physiological effect.
    I. Title. II. Title: Essential fatty acids in health and disease.
    [DNLM: 1. Fatty Acids, Essential—metabolism. 2. Diet. 3. Dietary Fats—metabolism.
    4. Health Behavior. 5. Nervous System—metabolism. QU 90 L421f 2010]
    QP752.F35L39 2010
    612.3'97—dc22
                                                                2009020397
                                                                CIP

A British Cataloging-in-Publication record for this book
is available from the British Library.

Visit our Web site: http://rutgerspress.rutgers.edu

Manufactured in the United States of America

CONTENTS

PART ONE
Nutritional, Chemical, and Physiological Properties of Dietary Fats

PART TWO
Dietary Fats in Health and Disease

PART THREE

Influence of Diet on Overall Health

FIGURES

TABLES

PREFACE

I had the good fortune to collaborate with several scientists on a wide range of research topics in the early 1980s, when I worked in the laboratory of Gerald Cohen at Mount Sinai School of Medicine in New York. My research focus was lipid peroxidation and the oxidative destruction of polyunsaturated fatty acids in cell membranes and its effects on aging, cancer, and inflammation. Free radicals are believed to be responsible for many cancers, and lipid peroxidation is thought to be one of the ways that free radicals are mediating carcinogenesis, as well as promoting various diseases of aging. There was evidence that dietary polyunsaturated oils were promoting some cancers.

Polyunsaturated fatty acids are considered to be essential dietary constituents because we are unable to make them from other nutrients and they have vital functions in numerous physiological systems in the human body. Sune Bergström, Bengt Samuelsson, and John Vane shared the Nobel Prize for Physiology or Medicine in 1982 for their discoveries related to metabolism of polyunsaturated fatty acids to powerful bioactive substances known collectively as eicosanoids. The scientific community was constantly discovering new lipid messengers and new roles for those that were already known. Eicosanoids were affecting nearly every system in the body, often in ways that were beneficial, but sometimes their actions were detrimental. The distinctions between omega-3 polyunsaturated fatty acids found in fish oils and omega-6 polyunsaturated fatty acids found in vegetable oils were beginning to unfold, and eicosanoids formed from them accounted for many observed differences in physiological responses to dietary fats from these disparate sources.

As I got involved in research projects related to inflammation, cancer, and epilepsy, I began to see that eicosanoids derived from essential fatty acids had powerful effects, and that manipulation of dietary fats that are metabolized to eicosanoids or not involved in their production could have profound influences on the course of several diseases. The polyunsaturated fatty acids nurtured a fascination and wonder for me, perhaps because of the complex ways in which they were affecting so many aspects of health and disease.

There was also considerable interest in saturated fats and cholesterol at that time. The results from two large-scale studies of the effects of dietary intervention and behavior modification on serum cholesterol and heart disease were emerging by 1985. The Multiple-Risk Factor Intervention Trial (MRFIT) and the Lipid Research Clinics Coronary Primary Prevention Trial each involved several thousand participants and hundreds of physicians and researchers. An aim of each study was to get people to reduce their consumption of saturated fats and cholesterol, which had been associated with heart disease. Michael Brown and Joseph Goldstein received the Nobel Prize in Physiology or Medicine in 1985 for their work on genetic causes for high levels of cholesterol in the blood that can lead to heart disease at a young age. Genetic factors were emerging as a major contributor to the risk of heart disease. Indeed, genetic factors seemed more important than dietary fats and cholesterol.

There was widespread belief that unsaturated vegetable oils were much healthier in the diet than saturated fats, particularly fats from animals that also contained cholesterol. I was involved in research that was showing that saturated fats were generally benign, whereas dietary polyunsaturated vegetable oils were susceptible to lipid peroxidation, which was associated with many of the common maladies of modern civilization. By the 1990s lipid peroxidation in low-density lipoprotein (LDL) was found to promote atherosclerosis and consequent heart disease, yet the doctrine of saturated fats and cholesterol causing heart disease persisted. Certainly very high levels of cholesterol in the blood were cause for serious concern. However, blanket recommendations for less saturated fat in the diet and implications that polyunsaturated oils were acceptable baffled me.

I was also astonished to see the popular health and nutrition literature condemning tropical oils, such as palm and coconut oils, because they contained high levels of saturated fats. I had never seen any studies that indicated consumption of these oils increased the incidence of any disease. However, significant pressure was placed on the food industry to remove tropical oils from their products, which resulted in their replacement with partially hydrogenated vegetable oils. The latter contained levels of saturated fats as high as those in palm oil but also had high levels of trans fats, which were known to be at least as bad as saturated fats with respect to serum cholesterol. The recommendations to replace tropical oils and butter with margarine and partially hydrogenated vegetable oils did not have any scientific evidence to support it. I was disturbed by the direction health recommendations were going, since polyunsaturated and trans fats were likely doing more harm than the saturated fats they were replacing.

I started writing down my concerns in the 1990s, but other commitments prevented me from bringing those writings to fruition. I followed the literature on polyunsaturated fatty acids regarding their various functions and health consequences. A few years ago I decided to begin writing this book in earnest. It required delving into a vast amount of literature on dietary fats and health, with certain leads causing diversions that were sometimes quite rewarding and resulted in expansion of the original scope of the book. A major underlying premise for the book was to inform readers about the potential adverse effects that dietary polyunsaturated oils could have, even though they are a source of essential fatty acids. The popular nutrition and health press was beginning to waiver on whether polyunsaturated oils were necessarily healthier and shifted to favor monounsaturated oils. However, I had great difficulty justifying an easement on the condemnation of dietary saturated fats. After all, the idea that saturated fats were a scourge on human health and that they increased the risk of heart disease had been ingrained in the public consciousness for more than half a century, even though scientific proof was lacking.

As the book progressed, I began to see scientific evidence that there were flaws in the low-saturated-fat and low-fat diet recommendations. Carbohydrates, especially sugars, were becoming the new plague on human health, and many health conditions that were formerly blamed on saturated fats could now be attributed to sugars. The saturated fat–cholesterol relationship was based on a simple premise that was easy to understand: saturated fats in the diet tend to raise serum cholesterol, whereas polyunsaturated oils tend to lower serum cholesterol. How polyunsaturated fatty acids influence the course of many diseases is more complicated, but far more important with regard to overall health. The inverse relationship between dietary fats and carbohydrates adds to the complexity. The role of dietary fats in health is not a simple matter, and I hope that you, the reader, will bear with me as I attempt to explain the scientific foundations for their complicated interactions in an understandable way. I believe that the arguments I present should persuade people to reconsider the dietary guidelines that were based on our earlier naivety.

The final version of the manuscript for this book is quite different from early drafts, and I have many people to thank for that transformation. I thank Lester Mayers, director of sports medicine at Pace University, Pleasantville Campus, and David Spierer of the Sports Science Department at Long Island University, Brooklyn Campus, for reading an entire early draft of the manuscript and offering helpful suggestions. As a reader selected by Rutgers University Press, Joseph Dixon of the Department of Nutritional Sciences, Rutgers University, rendered a detailed analysis of the manuscript that resulted in major improvements in the overall content of the book. Stuart E. Alleyne deserves much credit for his technical assistance in making many of the figures more presentable. I am indebted to Stuart Fishelson, a friend and colleague at Long Island University, for his constant encouragement to write this book and his assistance, along with Fred Farhadi, with the cover design. I am especially grateful to Doreen Valentine, acquisitions editor for Rutgers University Press, who was willing to consider the manuscript and offered extensive advice and suggestions in developing the original material into a more readable and significant work. Finally, I thank my wife, Alice Hendrickson, for her patience, understanding, and encouragement through all of this.

Glen D. Lawrence
Department of Chemistry and Biochemistry
Long Island University
Brooklyn, NY

ABBREVIATIONS

2-AG	2-arachidonylglycerol
5-HIAA	5-hydroxyindoleacetic acid
ADHD	attention deficit/hyperactivity disorder
ADP/ATP	adenosine diphosphate/adenosine triphosphate
AGE	advanced glycosylation end products
AIDS	acquired immune deficiency syndrome
ALS	amyotrophic lateral sclerosis
ApoA-I	apolipoprotein A-I
ApoE	apolipoprotein E
APP	amyloid precursor protein
BDNF	brain-derived neurotrophic factor
BMI	body mass index
CAM	cell adhesion molecule
CB1	cannabinoid 1 (receptor)
CLA	conjugated linoleic acid
COX	cyclooxygenase
CRP	C-reactive protein
cyclic AMP	cyclic adenosine monophosphate
DHA	docosahexaenoic acid
DHGLA	dihomo-gamma-linolenic acid
DNA	deoxyribonucleic acid
EP1–4	E prostaglandin receptor 1–4
EPA	eicosapentaenoic acid
FDA	United States Food and Drug Administration
GABA	gamma-aminobutyric acid
GLUT4	glucose transporter 4
HDL	high-density lipoprotein
HETE	hydroxyeicosatetraenoic acid
HMGCoA	3-hydroxy-3-methylglutaryl coenzyme A
IDL	intermediate-density lipoprotein
LCPUFA	long-chain polyunsaturated fatty acid
LDL	low-density lipoprotein
L-DOPA	L-dihydroxyphenylalanine

LOX	lipoxygenase
LTA_4–LTE_4	leukotriene A_4, B_4, C_4, D_4, E_4
LTP	long-term potentiation
MPP^+	N-methyl-4-phenylpyridinium ion
MPTP	N-methyl-4-phenyltetrahydropyridine
MS	multiple sclerosis
NMDA	N-methyl-D-aspartate (receptors)
NPD1	neuroprotectin D1
NSAID	nonsteroidal anti-inflammatory drug
PAF	platelet activating factor
PGE_2	prostaglandin E_2
PGH_2	prostaglandin H_2
PGI_2	prostacyclin or prostaglandin I_2
PLA_2	phospholipase A_2
PUFA	polyunsaturated fatty acid
SOD	superoxide dismutase
SREBP	sterol regulatory element binding protein
THC	tetrahydrocannabinol
TXA_2/TXA_3	thromboxane A_2/thromboxane A_3
VEGF	vascular endothelial growth factor
VLDL	very low-density lipoprotein

PART ONE

Nutritional, Chemical, and Physiological Properties of Dietary Fats

1

The Dietary Fat Doctrine

In the early twentieth century the major concern was whether diets provided sufficient calories and nutrients to ward off infections and disease. When antibiotics were introduced in the 1930s, those with access to these lifesaving drugs had longer and more productive lives. As a result of pushing the average life span ever higher, diseases of aging became commonplace. By the 1930s heart disease had become the leading cause of death in the United States, but the health-care community and the general public had other matters on their minds in the decade leading up to World War II.

In the late 1940s fat and cholesterol began to receive an extraordinary amount of attention in terms of funding for studies of the effects of dietary fats on health. A large number of participants were recruited for those studies, and a multitude of scientific publications were generated. It would seem that with all this effort we should now have a grasp of what dietary manipulations are needed to improve our health and longevity. Our understanding of the biochemical and physiological effects of dietary fats has advanced tremendously as a result of this onslaught of research, but where this knowledge leads us, in terms of sound dietary recommendations, is not so clear. Certain fats may improve one health problem but work against another.

Saturated fats are resoundingly denounced because they raise total serum cholesterol and low-density lipoprotein (LDL or LDL-cholesterol). There seems to be a consensus that this is bad. On the other hand, polyunsaturated fats lower serum cholesterol and LDL, and consequently are considered good. The stated effects that these two classes of fats have on total serum cholesterol and LDL-cholesterol are accurate, although their classification as bad and good are judgmental and apply primarily to serum cholesterol and LDL. If one looks at the effect these two types of fat have on cancers, the result is often, but not always, reversed. When it comes to arthritis or inflammation, there is a general consensus that the omega-6 polyunsaturated vegetable oils are bad, whereas omega-3 fish oils are better, and the saturated fats are most favorable. The influence of saturated and polyunsaturated fats on cancers, arthritis, and inflammation are less often mentioned in the popular health media, although the effects of omega-6 versus omega-3 fatty acids on inflammation are beginning to get more recognition.

Fats come in many forms: saturated, monounsaturated, polyunsaturated, omega-3, omega-6, partially hydrogenated, and trans fats. These terms are tossed around in health magazines, health sections of newspapers, on television and radio, and in daily conversations, but how well do we understand what it all means? Many people recognize the terms, but they are at a loss to identify which foods contain them. The nutrition facts section of a packaged food label gives the total amount of fat in a standard serving of that food with a breakdown of saturated fat, trans fat, and cholesterol, but is this giving sufficient information to make an informed decision about whether one product is better or worse than another? Labels for foods containing vegetable oils and other forms of fat may give the relative amounts of monounsaturated and polyunsaturated fats, but providing information on these categories is voluntary. Rarely is there an indication of the proportion of omega-3 and omega-6 polyunsaturated fatty acids on a product label. Even if such additional information were provided, would it make the consumer's decisions any easier?

It is easy to understand why there is so much confusion. These fats are not behaving in a consistent way over the full spectrum of all our ailments. This brings to mind the adage Feed a cold and starve a fever—or is it Feed a fever and starve a cold? When the scientific or biological basis for the saying is not understood, it is easy to forget which way it goes. The aim of this book is to inform the reader about how fats are influencing a variety of diseases, disorders, or imbalances so that reasonable dietary decisions can be made. When it comes to fats, such decisions are not as simple as one size fits all.

Roles of Fat

The physiological ability to store excess energy or caloric intake in the form of fat is a trait that confers numerous advantages across the biological spectrum. Fat is the most efficient way for living organisms to store excess energy. It provides more than twice as much energy as carbohydrate or protein: 9 kilocalories (kcal) per gram for fat versus 4 kcal per gram for carbohydrate and protein. Fat also excludes water, which can account for much of the mass of stored glycogen, the major starch and storage form of glucose in the human body. Fat can be formed from carbohydrate (sugars and starch) or protein, but glucose, the life-sustaining sugar in our circulation, cannot be formed from fat. Although many organs of the body can utilize fat for energy, the brain gets its energy almost exclusively from glucose. Special types of fats, specifically essential fatty acids, are required for brain development in the fetus, the newborn, and children, and are necessary for maintenance of proper brain function in adults.

What makes certain fatty acids essential will be discussed in more detail in chapter 3, although it will suffice to say here that the common dietary polyunsaturated fatty acids are essential fatty acids. Fatty acids are the main components of fats and of many other substances in nature. The distinctions between fats and fatty acids will be described in more detail in the next chapter. Polyunsaturated fatty acids are found predominantly in vegetable oils and fish oils. Mammals and humans are unable to make these fatty acids, so they must be supplied in the diet, much like vitamins. The polyunsaturated fatty acids are metabolized to a wide range of potent bioactive substances that have profound impacts on the physiology of the human (and animal) body.

Polyunsaturated fatty acids are also susceptible to chemical (nonenzymatic) oxidation, which leads to formation of numerous toxic products. These toxic products can alter DNA (deoxyribonucleic acid), the genetic material that dictates the fate of a cell and, perhaps, the organism of which that cell is a part. Certain mutations in DNA can lead to cancer, which can have dire consequences on the body as a whole. In other cases the alteration results in programmed death (apoptosis) of the affected cell, and the greater organism is spared. The factors that determine the fate of the cell in question are numerous and not completely understood, so it should not be surprising that the types of dietary fats ingested can have variable effects.

One of the essential polyunsaturated fatty acids is arachidonic acid. When we bleed from a cut, arachidonic acid is metabolized to a substance known as thromboxane A_2 that initiates a series of events to form a clot and prevent us from bleeding to death. The same thromboxane A_2 can trigger premature clotting in the coronary arteries to cause a heart attack. Other metabolites of arachidonic acid whip the immune system into action to ward off an infection, but these same metabolites may conspire to turn our immune systems against us when we are afflicted with an autoimmune disease. The polyunsaturated fatty acids are not alone in all these actions; they are helped by a wide array of protein cytokines and other biochemical factors.

Arachidonic acid is not the only polyunsaturated fatty acid involved in the myriad of complex biochemical and physiological events, but in a large majority of people it is the most common long-chain polyunsaturated fatty acid that undergoes metabolism to a wide array of biologically active substances that are collectively known as eicosanoids. Some of the other important polyunsaturated and essential fatty acids are known by their common abbreviations, EPA (eicosapentaenoic acid) and DHA (docosahexaenoic acid). These two may be better known and recognized as omega-3 fatty acids, whereas arachidonic acid is an omega-6 fatty acid (these distinctions are covered in the next chapter).

The functions of omega-3 fatty acids in the body can be quite different from those of arachidonic acid, but in some cases EPA competes with arachidonic acid as the raw material for metabolism to biologically active eicosanoids. The course of many physiological processes may depend on which of these fatty acids is in greater abundance as a result of their dietary intake. Evidence will be presented in many cases that the relative availability of omega-3 fatty acids compared with omega-6 fatty acids can influence the course of a disease.

The polyunsaturated fatty acids are classified as essential because our body needs them, but as we learn more about them, it is becoming clear that they have potential to cause harm. It is necessary to make distinctions between omega-3 and omega-6 polyunsaturated fatty acids in terms of their impact on several physiological systems in the body.

The Diverse Nature of Dietary Fats

Saturated and monounsaturated fatty acids are common in animal fats, such as meat and dairy products. These fatty acids can be made in the human body (as well as in animals), typically from excess carbohydrate (starches and sugars), and consequently are not classified as essential. Monounsaturated fatty acids are also abundant in certain oils derived from plants, notably olive, avocado, almond, canola, peanut, and palm oils. Monounsaturated oils are touted as the most favorable in the diet because they do not raise cholesterol as much as saturated fats do and they are less susceptible to spontaneous oxidation (decomposition) than the polyunsaturated fatty acids. In addition, they are not converted to the potent bioactive compounds that can disturb the balance of various systems in the body, which often happens with polyunsaturated fatty acids.

Many vegetable oils have an abundance of the omega-6 polyunsaturated fatty acid known as linoleic acid. It is converted in the body to arachidonic acid, which is involved in a multitude of physiological effects, some of which I have already described. The profile of fatty acids in various vegetable oils depends on many factors, such as the climate in which the plant source is grown, the genetic strain or variety of the plant, and the kinds of treatment the extracted oils receive during processing. Plants grown in colder climates tend to have more of the polyunsaturated linoleic acid in their triglycerides (storage fats), whereas those grown in warmer climates have more of the saturated palmitic acid. The agricultural industry has used sundry methods to obtain plant varieties that produce storage fats with distinct profiles of fatty acids. This can be done by genetic engineering but is usually accomplished through genetic selection and plant breeding.

The chemical breakdown products of polyunsaturated fatty acids, in general, are not healthy, which makes many vegetable oils undesirable for frying. Foods containing relatively large amounts of polyunsaturated oils often do not have a long shelf life because of their tendency to become rancid unless antioxidants or other preservatives are added. The food industry circumvented the problem of decomposition by hydrogenating vegetable oils, which converts polyunsaturated fatty acids to saturated and monounsaturated fatty acids. This chemical processing results in fats with longer shelf life and greater stability at high temperature, but it produces trans fats. Trans fats behave like saturated fats in terms of raising cholesterol and LDL, but they can also lower HDL-cholesterol and are suspected of having other undesirable health effects.

Manufacturers of processed foods are required to indicate how much trans fat their foods contain, although they can declare zero grams of trans fat if the product has less than one-half gram *per serving*. In the case of margarine, a standard serving size is considered to be one tablespoon or about 14 grams, meaning it can contain as much as 4 percent trans fat but still claim zero trans fat on the label. New York City became the first major city in the United States to prohibit the use of trans fats in restaurants. Many large restaurant chains needed to change their ingredients to meet the July 2008 deadline for the New York City ban on trans fat. The American Heart Association expressed concern that restaurants would merely switch from trans fats to natural cooking fats that are high in saturated fats, such as palm oil (MSNBC 2006). The fear of switching to saturated fats is based on their tendency to raise serum cholesterol, but saturated fats have fewer adverse effects than trans fats with respect to serum lipid profile and are preferable to polyunsaturated oils in terms of chemical stability.

Another important consideration when it comes to dietary sources of polyunsaturated fatty acids is to get a balance of omega-3 and omega-6 varieties. There is much concern that the ratio of these two series of essential fatty acids in the American diet has shifted dramatically in favor of omega-6 fatty acids. This is due primarily to modern agricultural and food-processing practices that favor a few varieties of grains for producing edible oils. There is much speculation, with a fair amount of scientific evidence to support it, that these large changes in the profile of essential fatty acids in the diet are responsible for many maladies of our modern civilization (Simopoulos 2002a). The omega-3 fatty acids are abundant in fish oils but are also found in varying proportions in some vegetable oils, such as flaxseed, walnut, canola, and soy oils, as well as in wheat germ. (The fatty acid composition of several dietary fats and oils are given in appendix B.)

Nature and economics rarely make it easy for us to decide what fats and oils are going to be best for our health. Olive oil tends to be significantly more expensive than most other vegetable oils, particularly in North America. Canola oil has a substantial amount of omega-3 polyunsaturated fatty acids, which is good from a health perspective, but those fatty acids decompose easily at high temperatures, making the oil less desirable for frying foods. Several years ago palm oil was condemned for having high levels of saturated fatty acids (about 50 percent) that could raise serum cholesterol and thereby increase the risk of heart disease. The claim regarding risk for heart disease was unsupported, but undue pressure from certain advocates forced food processors to switch from palm oil to partially hydrogenated vegetable oil in their formulations. That replacement probably did more harm than good, in view of the adverse effects of trans fats that were present in the hydrogenated oils at that time as well as the greater abundance of saturated fatty acids in hydrogenated oil compared with palm oil.

How dietary fats influence health and disease is complex. The decisions we need to make regarding what fats would be best in the diet are not simple. Some of the complications arising from polyunsaturated fatty acids have been mentioned. Monounsaturated fatty acids have fewer complexities, but dietary sources are composed of more than one type of fatty acid. The saturated fats have been condemned, but the evidence against them is not convincing.

Origin of Dietary Recommendations to Lower Cholesterol

By the mid-nineteenth century the chemical constituents of living matter were being characterized. Julius Vogel noted that cholesterol was a component of atherosclerotic plaque in his 1847 textbook *The Pathological Anatomy of the Human Body* (Kritchevsky 1998). In 1854 Rudolf Virchow proposed that athero-sclerosis was caused by an inflammatory response to injury of the arterial wall (McKinnon et al. 2006). It took nearly 150 years for scientists to unravel the biochemical processes of atherogenesis (formation of atherosclerotic lesions). Today it is well established that cholesterol is deposited in the lining of the arteries, particularly around the heart, as a result of immune cells clearing oxidized low-density lipoprotein (oxidized LDL) from the circulation. Nearly 50 percent of the mass of LDL is composed of cholesterol, and LDL seems to be the lipoprotein that is most prone to spontaneous oxidation.

In the early 1900s Nikolai Anitschkow showed that rabbits develop athero-sclerotic lesions when they are fed high-cholesterol, high-fat diets (Steinberg 2004). This work was dismissed by many scientists as an irrelevant model for human disease because rabbits are strict herbivores and would not normally consume cholesterol or high-fat diets. Similar studies in dogs and rats did not produce atherosclerosis. Nonetheless, it seemed reasonable to assume that atherosclerotic lesions were caused by having high levels of fat and cholesterol in the diet and subsequently in the blood. Virchow's idea that atherosclerosis was caused by inflammation in the walls of the arteries fell by the wayside in the early twentieth century but has been revived in recent years as we have learned more about the processes involved in atherogenesis.

During the 1950s several groups of researchers studied the effects of various types of dietary fat on serum cholesterol. The subjects in one study were residents of a large mental institution in New England, where all participants were housed and ate all meals in an isolated ward (Hegsted et al. 1965). Thirty-six different diets were carefully prepared to contain specific types and amounts of saturated, monounsaturated, and polyunsaturated fats and oils. A control diet was designed to resemble a standard American diet at the time. Subjects received test and control diets alternately for four-week periods in the first year, and received the control diet less frequently after the first year. The initial mean serum cholesterol level for men on the standard house diet (before receiving the control diet) was 225 milligrams per deciliter (mg/dL), which rose to 250 mg/dL when they consumed the special control diet.

Over time and several four-week dietary variations, the control diet produced a mean serum cholesterol level of 220 mg/dL, which means there was more than 10 percent difference in response to the same diet at different times. Coconut oil and butterfat, which have an abundance of shorter-chain (fewer than sixteen carbons) saturated fatty acids, raised serum cholesterol levels. However, cocoa butter, which is highly saturated but contains medium-chain fatty acids (sixteen or eighteen carbons in the chain), had little effect on serum cholesterol relative to the control diet. Polyunsaturated vegetable oils produced the largest decreases in serum cholesterol levels, while olive oil, which is high in monounsaturated fatty acids but also contains polyunsaturated fatty acids, caused moderate lowering of serum cholesterol.

Other studies involved patients or volunteers in hospital wards where diets were controlled and subjects were monitored for periods of several weeks (Bronte-Stewart et al. 1956; Keys 1957; Malmros and Wigand 1957). The general conclusion drawn from these studies was that dietary polyunsaturated oils lower serum cholesterol and saturated fats raise serum cholesterol, even though longer-chain saturated fats, similar to those found in most meats, had little effect compared with a standard or control diet. Today we have a better understanding of the mechanisms by which polyunsaturated fatty acids affect gene expression and alter the abundance of proteins involved in cholesterol metabolism.

Extraordinary emphasis was placed on these findings and their interpretations in view of the abundance of cholesterol in atherosclerotic plaque and little or no understanding of how it got there. A national campaign was launched by the American Heart Association to educate the public on the benefits of decreasing consumption of saturated fats and cholesterol, particularly from meat, eggs, and dairy products, and to use instead polyunsaturated oils as much as possible, mainly from vegetable sources (Kritchevsky 1998). Epidemiological studies helped to reinforce the saturated fat–cholesterol hypothesis.

The Epidemiology of Heart Disease

The study that probably had the greatest impact on assigning risk factors for heart disease was the Framingham Heart Study. This study set out in 1948 to monitor the health, eating, social, and behavioral characteristics of more than 5,200 residents of Framingham, Massachusetts (a city of about 28,000), who were between the ages of thirty and fifty-nine years. After several years of monitoring, the researchers reported a few factors that increased a person's risk for developing heart disease, which included cigarette smoking, high blood pressure, electrocardiogram abnormalities (arrhythmia), and high blood cholesterol (Kannel et al. 1961).

The six-year follow-up report for the Framingham study noted a dramatic difference in the number of new cases of heart disease (defined as sudden death attributed to no other causes, myocardial infarction, or angina pectoris) when comparing subjects with low (less than 210 mg/dL), intermediate (210 to 244 mg/dL), and high (greater than 244 mg/dL) serum cholesterol (Kannel et al. 1961). Each of these ranges for cholesterol comprised one-third of the sample population at the start of the study. The incidence of heart disease for men in the high-cholesterol group was more than three times that of men with the lowest cholesterol (12 percent versus 3.5 percent). Although the incidence of new heart disease among men in the intermediate-cholesterol group (6.4 percent) was about 80 percent higher than for those with low cholesterol, the difference was not statistically significant. The relationship between serum cholesterol and heart disease among women showed a similar trend, but the incidence of heart disease was much less for women. The difference in the number of new heart disease cases between groups were less striking as the study went on and more people reached older ages.

The Framingham Heart Study measured several medical diagnostic parameters related to heart disease, but for a relatively homogeneous population in a single location. A much more ambitious study was initiated by Ancel Keys from the University of Minnesota. He designed a study that collected similar medical data, as well as dietary information, on sixteen cohorts in seven different countries. The countries included Finland, Greece, Italy, Japan, the Netherlands, the former Yugoslavia (Serbia and Croatia), and the United States. The only American cohort was made up of railroad workers in the northwestern United States. Keys published the results of ten years of follow-up on all cohorts in a 381-page monograph (Keys 1980).

A major aim of the Seven Countries Study was to determine what medical, dietary, and behavioral (smoking) factors influenced the incidence of coronary heart disease, death from coronary heart disease, and death from all causes over a ten-year period. The two factors that proved most significant in predicting coronary deaths and all-cause deaths in the Seven Countries Study were age and high systolic blood pressure (Keys 1980, 321). However, systolic blood pressure posed a significantly elevated risk only for those participants in the highest 20 percent of blood pressure in their cohort.

Perhaps the most frequently cited statistic from the Seven Countries Study is the correlation between the ten-year coronary death rate and median serum cholesterol level for the various cohorts. There were a few anomalies in the graphic representation, notably the eastern Finland cohort with by far the highest rate of coronary deaths (681 per 10,000 men) and the Crete (Greece) cohort with zero coronary deaths in the ten years of the study. Interestingly, the western Finland cohort had less than half the rate of coronary deaths (250 per 10,000 men) compared with their eastern countrymen, but their mean serum cholesterol level was only about 10 percent lower—the two Finnish cohorts had the highest serum cholesterol of all cohorts (Keys 1980, 122). The coronary death rate for western Finland was slightly above the average for all cohorts, but it was substantially below that of the American railroad men (424 coronary deaths per 10,000 men).

There was no significant correlation between the median serum cholesterol level and death rate from all causes among the cohorts. In fact, there was a moderate negative correlation between serum cholesterol and noncoronary death rate, which would indicate that serum cholesterol levels may not be a good predictor of death from all causes. Indeed, large studies of the effect of dietary changes on serum cholesterol and incidence of heart disease or coronary deaths have shown that dietary manipulations to lower serum cholesterol do not affect the all-cause death rate (Lipid Research Clinics Study Group 1984; Multiple Risk Factor Intervention Trial Research Group 1982; WHO 1984).

Both the mean cohort serum cholesterol level and the ten-year age-adjusted coronary death rate were highly correlated with saturated fats in the diet. The incidence of new cases of coronary heart disease showed a weaker correlation with dietary saturated fats, as well as with dietary sucrose. The correlation between heart disease and dietary sucrose was attributed to an intercorrelation between sucrose and saturated fats in the diet. Today we have a better understanding of the influence of dietary sugar on cardiovascular disease. Excess sugar in the diet can increase serum triglycerides, which are a risk factor for heart disease (Fried and Rao 2003). But a high level of sugar in the diet has potential to do more in terms of influencing heart disease, as well as other health complications.

Keys noted that the all-causes death rate tended to rise only when serum cholesterol was above about 275 mg/dL (which is considered very high). Even for coronary deaths, there was no relationship to serum cholesterol below about 230 to 240 mg/dL (Keys 1980, 325). The early report on risk from the Framingham Heart Study divided participants into only three groups based on serum cholesterol. A serum cholesterol level of 244 mg/dL was used as the lower limit for the highest cholesterol group simply because one-third of the participants had levels of cholesterol higher than that. Consequently, the impression that heart disease risk becomes greatest when serum cholesterol approaches 240 mg/dL can be misleading. The much greater incidence of heart disease in the high-cholesterol group of the Framingham study was likely due to participants that had extremely elevated cholesterol (for example, above 275 mg/dL as found in the Seven Countries Study).

The Seven Countries Study data are valuable in that they show that ethnic and cultural factors may play an important role in the diet–heart disease connection. Although serum cholesterol correlated with ten-year coronary death rate within some cohorts, there was no significant correlation in others—notably the Mediterranean countries (Italy, Greece, and the former Yugoslavia). Ancel Keys was architect and avid promoter of the Mediterranean diet (Ancel and Margaret Keys 1975). He died in 2004 at the age of one hundred, a strong argument for his conviction.

The Framingham Study and the Seven Countries Study are presented here to show how cholesterol became the focus of the medical establishment. They show that dietary saturated fats raise serum cholesterol compared with unsaturated oils, but the level of serum cholesterol that will significantly shorten one's life span is not as clear. The chances of dying prematurely increases dramatically when serum cholesterol gets above about 250 mg/dL, but the correlation between serum cholesterol and death from all causes diminishes when cholesterol levels are below 240 mg/dL. The point that needs to be emphasized here is that although serum cholesterol levels are an important consideration regarding risk of heart disease, there is a level (about 240 mg/dL) below which the concern about serum cholesterol should diminish. There are many other health problems afflicting the American population, and greater consumption of dietary polyunsaturated oils that lower serum cholesterol may not be the best choice for dealing with those other health problems.

Other Factors to Consider Regarding Dietary Fats

There are several flaws in the logic of substituting polyunsaturated oils for saturated fats in the diet to increase life span. First, the majority of myocardial infarctions (heart attacks) occur with less than 50 percent occlusion of the arteries by atherosclerosis, so something other than blockage of the arteries by cholesterol buildup is triggering heart attacks (Smith 1996). This is not to say that severe atherosclerosis is no cause for concern—a majority of people with greater than 50 percent occlusion of the arteries will probably suffer some sort of heart disease. There are several factors, including destruction of atherosclerotic plaque, that cause heart attacks (see chapters 6 and 7). Polyunsaturated fatty acids are intrinsically involved in many of the physiological processes that contribute to heart disease.

A second reason for doubting the supposed benefits of polyunsaturated oils is that genetic predisposition is a major factor contributing to high levels of LDL in the blood and consequent risk for heart disease. If someone has elevated serum total cholesterol (greater than 240 or 250 mg/dL) and LDL-cholesterol (greater than 150 mg/dL), that person is at high risk for cardiovascular disease. If other risk factors are present, such as diabetes, high blood pressure, low HDL-cholesterol, or a cigarette habit, the National Cholesterol Education Program recommends the use of lipid-lowering drugs when much lower levels of total cholesterol and LDL-cholesterol are inherent (Grundy et al. 2004). There are numerous genetic factors that increase or reduce the relative risk for heart disease, or other life-threatening diseases for that matter. Levels of LDL in the blood are influenced mostly by genetic factors, although diet may have a moderate effect. Numerous genetic variants have been identified that cause elevated LDL-cholesterol (Goldstein and Brown 2001). Genetic variations in elements unrelated to LDL-cholesterol can also affect heart disease. Changes in dietary fats have relatively little influence on several of these genetically based disorders.

A third factor worth considering is that although polyunsaturated fatty acids can lower serum cholesterol, they are susceptible to oxidative degradation. If they are part of low-density lipoprotein particles when this happens, the oxidized LDL will be removed from the blood and deposited in the lining of arteries to promote atherosclerosis. Although saturated fats tend to increase the amount of circulating LDL, they are not susceptible to chemical oxidation. This conundrum makes it difficult to decide where to draw the line regarding relative amounts of saturated, monounsaturated, and polyunsaturated fats in the diet. A relative balance of these without a large excess of polyunsaturated oils would seem most prudent at this time.

Finally, dietary manipulations alone in large study populations typically do not influence the number of heart attacks or death from all causes (as mentioned above). The polyunsaturated fatty acids found in most vegetable oils are a double-edged sword. In addition to lowering serum cholesterol, they are metabolized to a wide range of highly potent bioactive substances, such as prostaglandins, thromboxanes, and leukotrienes to name a few, that evoke beneficial physiological actions throughout the body. However, they can get out of control and wreak havoc in many ways.

The high levels of omega-6 vegetable oils consumed in industrialized nations are being blamed for a wide range of diseases, including asthma (Black and Sharpe 1997), diabetes (Berry 1997), and inflammatory diseases such as rheumatoid arthritis, ulcerative colitis, and psoriasis (Heller et al. 1998). Cancer studies in laboratory animals have shown that vegetable oils promote tumors to a greater degree than fish oils or saturated fats (Braden and Carroll 1986). Vegetable oils may not be the answer to increased life span and improved quality of life.

Moving Forward

Polyunsaturated fatty acids cannot be lumped together as purely good or bad. Omega-3 polyunsaturated fatty acids are plentiful in fish oils, and omega-6 polyunsaturated fatty acids abound in most vegetable oils. When omega-3 and omega-6 polyunsaturated fatty acids are metabolized to potent hormone-like substances, they elicit subtle (and sometimes not so subtle) differences in physiological effects that can have a profound influence on health and well-being. There is increasing evidence that the ratio of omega-3 to omega-6 polyunsaturated fatty acids in the diet has an impact on brain development and neurological disorders as well.

The high levels of polyunsaturated fatty acids consumed by many people today are carrying out many important beneficial functions in the body, but they are also responsible for a lot of pain and agony. There can be many varied responses to dietary changes and drug intervention as a result of an individual's genetic makeup. The concerns about cholesterol have clouded our view regarding dietary and pharmaceutical approaches to our collective health and well-being. Obesity, metabolic syndrome, and diabetes are further complicating the picture. The cholesterol story is important, but we should not lose sight of sensible dietary practices out of fear of cholesterol. As our understanding advances with regard to the biochemical, physiological, and pharmacological processes involving dietary fats, particularly the essential fatty acids, it becomes necessary to step back and reassess the recommendations that evolved when there was a less complete knowledge of their effects.

2

Lipid Terminology, Structure, and Function

The purpose of this chapter is to give some of the basics regarding biochemical properties of lipids and the terminology used to describe them. It is intended to introduce readers to several classes of lipids and the subtle distinctions that make these classes different from one another. *Lipid* refers to oily or fatty biological substances that are not soluble in water but dissolve in other solvents, such as chloroform, alcohol, or hexane. The term lipid encompasses fatty acids, triglycerides, phospholipids, the fat-soluble vitamins (A, D, E, and K), steroids (cholesterol and many of its metabolic products), and a variety of less commonly discussed biological materials such as terpenes. *Fat* is often thought to be synonymous with lipid, but fats are technically considered to be the triglycerides and, in some circles, only those triglycerides that solidify at room temperature. Triglycerides that remain liquid at room temperature are considered to be oils, although a substance need not be a triglyceride to be classified as oil. These distinctions are often overlooked today.

The fact that lipids do not dissolve in water is one of their most notable features. Such chemical incompatibility is due to the fact that lipids are nonpolar molecules, while water is polar. Water molecules behave like tiny magnets, sticking to one another through positive and negative poles (like the north and south poles of a magnet). This is what causes water to bead up on a freshly waxed surface. Lipids lack these polar properties, behaving more like globs of grease and fat—they don't have magnetic personalities. In other words, "like dissolves like," or "oil and water don't mix."

The idea of polar versus nonpolar substances will be used frequently in this book, along with terms such as *hydrophilic* (water loving), *hydrophobic* (water fearing), and *lipophilic* (lipid loving). For most biological substances, "fearing" water and "loving" lipids amounts to the same biochemical behavior. Appendix A reviews the fundamentals of chemical bonding and polarity.

Fatty Acids

The simplest lipids are the fatty acids. Fatty acids are typically components of other classes of lipids, particularly triglycerides, phospholipids, and cholesteryl esters. As the simplest unit, fatty acids can be transferred between or removed from these other lipids, a process that is fundamental to a wide range of metabolic actions and physiological functions. The fatty acids themselves, particularly the polyunsaturated fatty acids, are the major focus of this book, so it is important to get to know several of these by name. It will also be important to associate the names with some of their chemical or structural characteristics, such as saturated versus unsaturated, and omega-3 versus omega-6 polyunsaturated fatty acids.

There are four different fatty acids illustrated in figure 2.1. Each of them contains eighteen carbon atoms (as well as hydrogen and oxygen), but they differ in whether or not they have carbon-to-carbon double bonds, the number of double bonds, and the configuration of the double bond (cis or trans). When all the carbon atoms are connected by single bonds, the fatty acid is saturated (it has the maximum number of hydrogen atoms, hence it is saturated with hydrogen). In order to form a carbon-to-carbon double bond, a hydrogen atom must be removed from each carbon forming the double bond, making the molecule unsaturated with hydrogen. When there is one carbon-to-carbon double bond, the molecule is monounsaturated, whereas polyunsaturated fatty acids have more than one double bond.

Stearic acid (far left side in figure 2.1) is a common saturated fatty acid found in foods, but palmitic acid (with sixteen carbon atoms) is the predominant saturated fatty acid in most foods, as well as in the human body. Oleic acid (left-center in figure 2.1) is a major monounsaturated fatty acid that has the cis configuration for the double bond, which gives rise to a bend in the carbon chain. The significance of the bend in cis unsaturated fatty acids will be discussed below. Oleic acid is the major fatty acid found in olive oil, palm oil, and canola oil, which are promoted as monounsaturated oils. Elaidic acid (far right side in figure 2.1) has one carbon-to-carbon double bond but has a trans configuration that imparts a slight kink in the carbon chain. However, the carbon chain of elaidic acid is essentially straight, like that of a saturated fatty acid (such as stearic acid).

If there are two or more double bonds, the fatty acid is polyunsaturated, such as linoleic acid (right-center in figure 2.1). The second double bond puts an additional bend in the carbon chain. Linoleic acid is the most common polyunsaturated fatty acid found in most vegetable oils. It is also an omega-6 fatty acid because there is a double bond on the sixth carbon from the methyl end of the chain (bottom of the diagram), or the opposite end from the acid group. Alpha-linolenic acid (or linolenic acid) is less common in vegetable oils but is a significant omega-3 polyunsaturated fatty acid in some oils, such as flaxseed

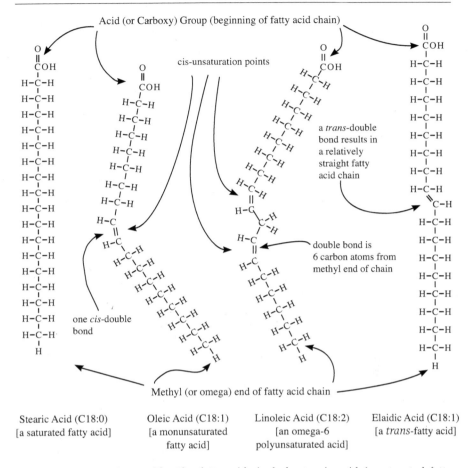

FIGURE 2.1. Four fatty acids. The fatty acids include stearic acid (a saturated fatty acid), elaidic acid (a trans-unsaturated fatty acid), oleic acid (a monounsaturated fatty acid), and linoleic acid (an omega-6 polyunsaturated fatty acid). The naturally occurring unsaturated fatty acids from plant sources contain cis double bonds at the unsaturation points. Notice the curvature of the cis unsaturated fatty acid chains compared with the linearity of the trans-unsaturated fatty acid.

oil, canola oil, soy oil, and wheat germ oil. Linolenic acid is shown in figure 2.2. Omega-3 is used to denote the fact that there is a double bond on the third carbon from the methyl end of the fatty acid chain. It is easy to confuse the names *linoleic* acid and *linolenic* acid, but the former is an omega-6 fatty acid and the latter an omega-3 fatty acid. The characteristics of many fatty acids, their physical properties, and abundance in various food sources are shown in appendix B.

There are several properties that determine the behavior of fatty acids and other lipids that contain fatty acids: the overall length of the carbon chain, the number of carbon-to-carbon double bonds, and the configuration of the double bonds (cis or trans). Shorter-chain fatty acids (fewer carbon atoms per molecule) tend to be liquids, whereas longer-chain fatty acids (more carbon atoms) tend to be solids. There is a transition from liquid to solid at room temperature when the saturated fatty acid chain length reaches about twelve carbons (see appendix B for melting points).

The degree of unsaturation (the number of double bonds) will influence the physical properties of a fatty acid and the lipids containing that fatty acid. The more double bonds there are within a given chain length, the lower the melting point will be. Consequently, the melting points decrease from stearic acid (70°C) to oleic acid (13°C) to linoleic acid (−9°C) and finally to linolenic acid (−17°C). This means the more double bonds a fatty acid has, the more fluid (or liquid) it will be. Saturated fats tend to be solid at room temperature, whereas monounsaturated oils remain liquid to lower temperatures, and polyunsaturated fatty acids remain liquid well below the freezing point of water.

Another important consideration is whether the double bonds have cis or trans configurations. *Cis* means that both parts of the carbon chain extend from the same side of the double bond, as the left arm and left leg extend from the same side of the trunk of the body. Notice how the carbon chain extends from the same side of the double bond in oleic acid (figure 2.1). *Trans* means that the carbon chain extends from different sides of the double bond, as shown for elaidic acid in figure 2.1. This is analogous to the left arm and the right leg extending from different sides of the body. Most naturally occurring double bonds in fatty acids are cis, resulting in a bend in the fatty acid chain (see figure 2.1). When there is a trans double bond, the molecule remains relatively straight.

It should be easy to see that the saturated and trans unsaturated fatty acids shown at the left and right could stack together tightly. On the other hand, the cis unsaturated fatty acids with their bent chains would behave in a more disorderly fashion. We generally think of the saturated and trans unsaturated fatty acids as rigid or solid, whereas the cis unsaturated fatty acids are fluid; the more unsaturated, the more fluid they would be. These principles apply for other classes of lipids that contain fatty acids, such as triglycerides and phospholipids discussed below.

The trans-fatty acids are found almost exclusively in partially hydrogenated vegetable oils, such as margarine and shortening. Hydrogenation was first applied to vegetable oils by Wilhelm Normann in 1902; he combined hydrogen gas with polyunsaturated fatty acids in the presence of a catalyst to convert the polyunsaturated into saturated and monounsaturated fatty acids (Kirschenbauer 1960). This chemical process results in the formation of trans double bonds instead of the cis double bonds that occur naturally in vegetable oils.

Six of the common polyunsaturated fatty acids are shown in figure 2.2. The eighteen-carbon polyunsaturated fatty acids are shown on the left side. The polyunsaturated fatty acids in the upper half of this figure are classified as omega-6 because there is a double bond on the sixth carbon from the methyl end of the chain (compare linoleic acid in figure 2.1 and figure 2.2). The chemical structures in figure 2.2 are shown without the C for carbon atoms and H for hydrogen atoms used in figure 2.1. Linoleic acid and arachidonic acid are essential fatty acids because the body cannot make them from other nutrients, such as carbohydrates or protein, and they are needed to maintain a healthy state. Dihomo-gamma-linolenic acid is not universally considered an essential fatty acid, but it should be placed in this category because it can function as one. (The idea of essentiality is discussed in more detail in the next chapter.) The omega-3 polyunsaturated fatty acids are shown in the lower half of figure 2.2. These include linolenic acid, eicosapentaenoic acid (EPA), and docosahexaenoic acid (DHA). All the omega-3 polyunsaturated fatty acids are generally considered to be essential.

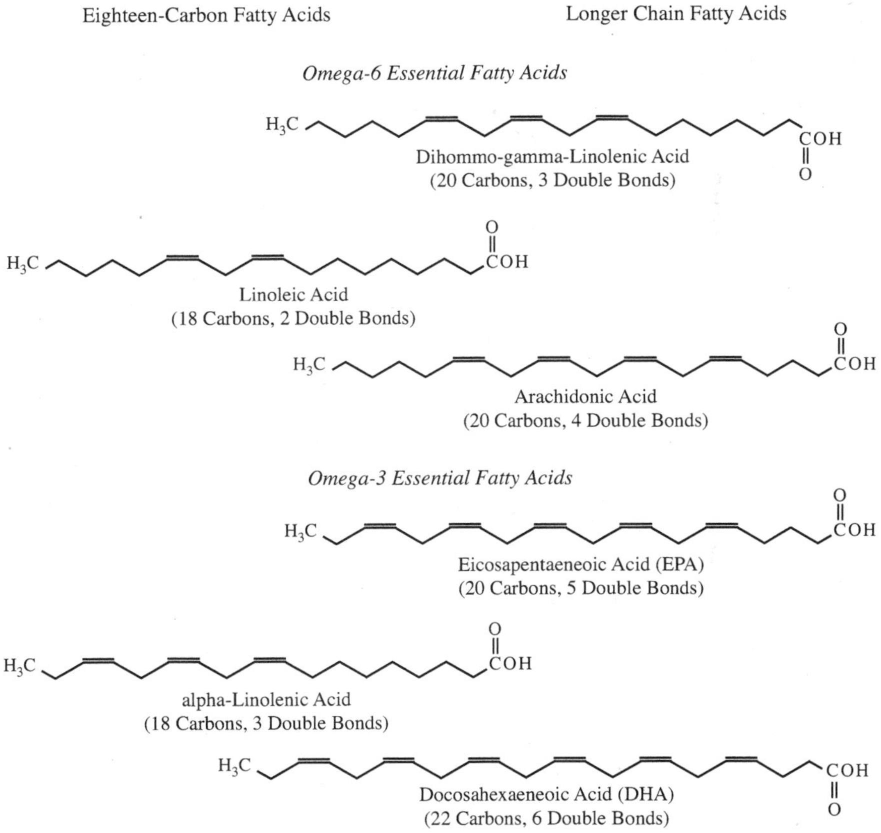

FIGURE 2.2. The essential fatty acids. All the essential fatty acids are polyunsaturated. They are divided into the omega-6 (*top half*) and omega-3 (*bottom half*) fatty acids, depending on whether the first double bond occurs at the third or sixth carbon atom of the carbon chain (counting from the methyl end of the molecule).

Storage Fats

It is more efficient to store energy in the form of triglycerides than in the form of carbohydrates or proteins. Our bodies have a limited capacity for storing glycogen (animal starch) as a glucose reserve, so when there is an abundance of carbohydrates in the diet, the excess will be converted to fat. There seems to be a virtually unlimited capacity to store excess energy as triglycerides in fat tissue, as exemplified by cases of extreme obesity in humans and some genetically manipulated animal species.

Although fatty acids are not particularly reactive, our metabolic machinery has evolved to keep the fatty acids attached to other molecules. Glycerol (also called glycerin) is a three-carbon molecule with an alcohol group (-OH) on each carbon. When fatty acids are attached to each alcohol of the glycerol, a triglyceride is formed. Another name for triglyceride is triacylglycerol, because there are three (tri) acyl (from acid) groups attached to the glycerol. The chemical structure of triglycerides is represented in figure 2.3. Notice that the same fatty acids shown in figure 2.1 have been used to form the triglycerides, but the symbols used to represent them have been modified, as they were for figure 2.2.

The triglyceride on the left in figure 2.3 is formed from unsaturated fatty acids. Empty spaces between the fatty acid chains allow some movement among the molecules (fluidity). The triglyceride on the right in figure 2.3 is composed of saturated and trans unsaturated fatty acids that tend to stack tightly together,

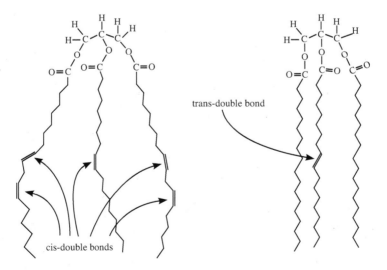

FIGURE 2.3. Two triglycerides. An unsaturated triglyceride (*left*) found in vegetable oils and a partially hydrogenated triglyceride (*right*) containing trans-unsaturated and saturated fatty acids after chemical hydrogenation.

leaving little space between molecules for movement (rigidity). Triglycerides containing trans fatty acids will be more solid at room temperature, as in margarine or shortening, compared with the polyunsaturated vegetable oils from which they are made.

Desaturation and Fate of Polyunsaturated Fatty Acids

All organisms are capable of synthesizing fatty acids from other nutrients. Specifically they are formed by breaking down carbohydrates (and possibly protein) and then building fatty acids by adding two carbon atoms at a time. Most organisms (except some bacteria) synthesize fatty acids this way, so virtually all the fatty acids consumed in a normal diet have an even number of carbons in the chain. Fatty acid chains with eighteen carbons are generally the most abundant in our diets, but there are also ample amounts of sixteen-, twenty-, and twenty-two-carbon fatty acids, and lesser amounts of those containing six, eight, ten, twelve, fourteen, and twenty-four carbons. Fatty acids with more than twenty-four carbon atoms are extremely rare.

Fatty acids are initially synthesized as saturated fatty acids, and then specific enzymes are responsible for desaturating them (placing double bonds at specific sites in the chain). In mammals (and humans) the enzymatic process stops when there are sixteen carbons in the chain. The sixteen-carbon saturated fatty acid, palmitic acid, is the most abundant fatty acid in most human tissues. In a separate synthetic step, the chain can be further elongated by adding two carbons at a time; it can also be desaturated. The latter is accomplished by enzymes known as desaturases that put double bonds at specific positions. Because genes contain the information for making enzymes (which are proteins), the type of desaturases any organism has will depend on the genetic makeup of that organism. Humans, and mammals in general, have a limited ability to synthesize polyunsaturated fatty acids.

Humans and other mammals have genes for desaturases that place double bonds at positions 5, 6, and 9 in the fatty acid chain (between C5 and C6, between C6 and C7, or between C9 and C10) (McGarry 2002, 704). These are named delta-5-, delta-6-, and delta-9-desaturase, respectively. Sometimes the Greek letter delta (Δ) is used to symbolize positions for double bonds. Humans and other mammals are not able to synthesize unsaturated fatty acids with a double bond beyond the delta-9 position (beyond C9 and C10 of the fatty acid chain).

Desaturation is not random, because desaturase enzymes have a preference for certain fatty acids; they select a fatty acid with a certain number of carbon atoms in the chain and a certain number of double bonds already present. The delta-6-desaturase has a preference for fatty acids with eighteen carbons and more than one existing double bond. Linoleic acid and linolenic acid are the most common polyunsaturated fatty acids in the diet that are desaturated by the delta-6-desaturase. The metabolism of these polyunsaturated fatty acids usually does not stop there. For example, two additional carbon atoms are added by a process known as elongation, and then another double bond is formed by the delta-5-desaturase enzyme (as shown in figure 2.4). The final product of this reaction sequence is arachidonic acid, which has twenty carbon atoms and four double bonds at positions 5, 8, 11, and 14 relative to the acid group. Arachidonic acid will enter into many of the discussions throughout this book in the context of essential fatty acids and eicosanoid metabolism.

Linolenic acid is metabolized to its corresponding twenty-carbon fatty acid, which has five double bonds at positions 5, 8, 11, 14, and 17. This is known as eicosapentaenoic acid or EPA for short. Linoleic and arachidonic acids are omega-6 polyunsaturated fatty acids because the double bonds are on the sixth carbon from the methyl end of the chain (see figure 2.1), whereas linolenic acid and EPA are omega-3 polyunsaturated fatty acids because there are three carbon atoms from the double bond to the methyl end of the chain (see figure 2.2). Another important omega-3 fatty acid is docosahexaenoic acid (DHA), a twenty-two-carbon fatty acid with six double bonds, which will be described in more detail in the next chapter.

Such desaturation and elongation usually does not happen to fatty acids that are destined for storage in the form of triglycerides. These long-chain fatty

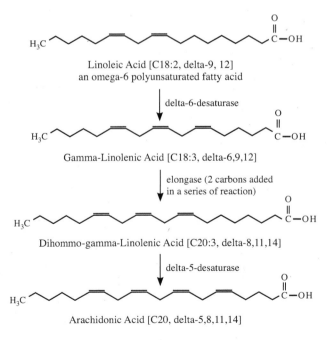

Linoleic Acid [C18:2, delta-9, 12]
an omega-6 polyunsaturated fatty acid

delta-6-desaturase

Gamma-Linolenic Acid [C18:3, delta-6,9,12]

elongase (2 carbons added
in a series of reaction)

Dihommo-gamma-Linolenic Acid [C20:3, delta-8,11,14]

delta-5-desaturase

Arachidonic Acid [C20, delta-5,8,11,14]

FIGURE 2.4.   Metabolism of linoleic acid to arachidonic acid. This involves initial desaturation at carbon 6, followed by elongation to increase the chain length by two carbons (through a series of enzyme-catalyzed reactions), and final desaturation at carbon 5.

acids with twenty or more carbon atoms and multiple double bonds (polyunsaturated), such as arachidonic acid, EPA, and DHA, are generally found in another class of lipid known as phosphatides or glycerophospholipids.

Polar Lipids

Phosphatides are known as polar lipids because they have a polar phosphate group attached to one position of the glycerol molecule. An additional chemical appendage is usually attached to the phosphate, such as choline, serine, ethanolamine, or inositol, giving rise to phosphatidyl choline, phosphatidyl serine, phosphatidyl ethanolamine, and phosphatidyl inositol, respectively. A typical phospholipid known as lecithin (phosphatidyl choline) is shown in figure 2.5. Notice that the phosphatides still have two fatty acids attached to the glycerol, making that portion of the molecule nonpolar.

Phospholipids make up the membranes of cells and are detergents or emulsifying agents used for the digestion of fats in the gastrointestinal (GI) tract. They are also components of the outer shell of lipoproteins, such as low density lipoprotein (LDL) and high density lipoprotein (HDL), which transport fats and other lipids through the bloodstream.

Sphingolipids are another class of polar lipid containing sphingosine, which has a long, eighteen-carbon chain with one double bond, an amino group for attaching one fatty acid (as an amide rather than an ester), and may have a phosphate attached to the terminal -OH group (see figure 2.6). Alternatively, there may be one sugar or several sugar units (oligosaccharide) attached to the terminal -OH group of sphingosine, forming what are known as cerebrosides or

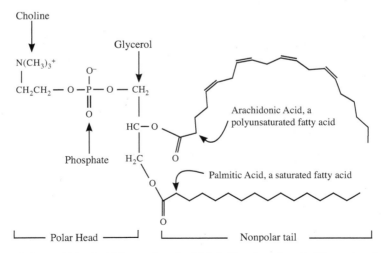

FIGURE 2.5. Lecithin. Lecithin, or phosphatidyl choline, is a phosphatide, a phospholipid, and a polar lipid.

gangliosides, respectively. When sugars or oligosaccharides are attached to the sphingosine, the resulting sphingolipids are also classified as glycolipids. The sugar groups constitute the polar, water-loving (hydrophilic) portion of these lipids in lieu of a phosphate group.

Oligosaccharides may also be attached to proteins, forming glycoproteins. The glycolipids and glycoproteins are found on the outer surface of cell membranes and give rise to antigenic determinants recognized by the immune system as self or nonself. The roles that sphingolipids play in cells are only beginning to be elucidated. Genetic disorders that result in the absence or deficiency of specific enzymes involved in degradation of these glycolipids can have severe health consequences, as seen in Tay-Sachs, Fabry's, and Gaucher's diseases, among others.

Membranes

Complex lipids such as phospholipids and glycolipids can have polar and nonpolar properties within the same molecule. *Amphipathic* (having strong

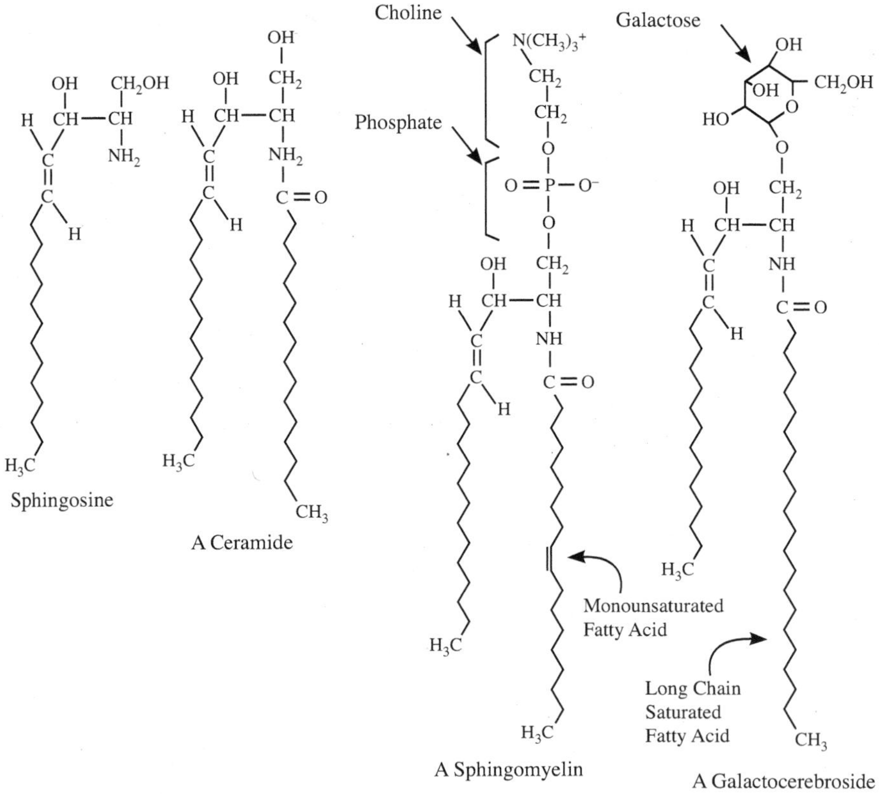

FIGURE 2.6.   Structures of sphingosine and some sphingolipids.

feelings for both sides—polar and nonpolar) is the technical term used to describe this property. The nonpolar portion consisting of the two fatty acids will shy away from water and mingle with the fatty acid portion of other phospholipids or triglycerides, while the polar phosphate or carbohydrate groups will tend to associate with water. The phospholipids are usually drawn schematically as a small circle to represent the polar phosphate end and two squiggly lines attached to represent the two attached fatty acid chains. These will spontaneously combine with one another in an aqueous environment to form a bilayer membrane as depicted in figure 2.7, with water on either side of the membrane; when the membrane is defining a cell, the water will be inside and outside the cell.

Cell membranes are not simply phospholipids forming a bilayer but are composed of many other components, such as vitamin E, cholesterol, and proteins. Vitamin E and cholesterol are also classified as lipids, but proteins are in a class of their own. There are many different kinds of proteins in a cell, including structural proteins, enzymes that catalyze biochemical or metabolic reactions, receptors, and transport proteins. All these proteins carry out specific functions in membranes. Cell membranes were once thought to be static, semi-permeable envelopes around cells, but it is now clear that they have dynamic functions, with all kinds of changes, generally catalyzed by enzymes, constantly taking place in the membrane. Transport proteins are another part of the dynamic, bringing nutrients into the cell and sending waste or useful products out of the cell through the membrane.

The nature of the fatty acids attached to the phospholipids dictates whether they are rigid (solid) or fluid (liquid). If there are long-chain saturated fatty acids attached, the phospholipids will be less fluid at body temperature. If a monounsaturated fatty acid is present, the fluidity increases; a polyunsaturated fatty acid (as shown in figure 2.5) will make membranes even more fluid. The shape of polyunsaturated fatty acids seen in figure 2.5 allows the phospholipids

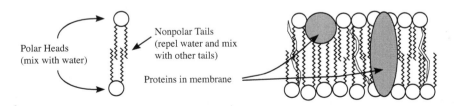

FIGURE 2.7.   Two phosphatides (*left*) and a phospholipid membrane (*right*) with several phosphatides and two proteins embedded in the membrane. The proteins may be exposed to one side of the membrane or pass completely through the membrane (left and right proteins, respectively). Each circle with two squiggly lines attached represents a phosphatide, such as lecithin (shown in figure 2.5) or a sphingolipid, such as a ceramide (shown in figure 2.6).

to roll or bounce off one another easily, whereas the straight chains of saturated or trans unsaturated fatty acids cause the phospholipids to stack more tightly together. Saturated fatty acids are often associated with sphingolipids, whereas long-chain polyunsaturated fatty acids are more often associated with phosphatides. Monounsaturated fatty acids are abundant in both.

One can think of the highly unsaturated lipids as behaving like beads or balls—if one fills a jar with these, they tend to flow over one another easily. On the other hand, the more saturated polar lipids would behave more like a jar filled with toothpicks packed tightly together that will not flow easily. Fluidity refers to how objects embedded in the membrane may be able to move around within the membrane relative to one another, much like passengers on a commuter train. If there is space between passengers, it is possible for them to move around within the train. However, if the car is crammed full of passengers with no space between them, it becomes a rigid mass of people with little possibility for movement through the car.

Cholesterol is incorporated into cell membranes for various purposes. The fluid mosaic model of membrane structure proposed that various membrane constituents are embedded in the bilayer lipid membrane (as in figure 2.7) and move around in this fluid shell, much like debris floating in a pond. It was thought that cholesterol would make the saturated phospholipids more fluid, while making the polyunsaturated phospholipids less fluid. This early model is evolving into something a little more complex as we begin to understand more about lipid interactions taking place in the membrane, as well as interactions with proteins that have a wide range of specific functions in the membrane. Because of their dynamic nature, it is essential that membranes remain fluid or pliable, but excess fluidity can be detrimental.

Cells in the body regulate the fluidity of membranes by controlling the fatty acids used in the phospholipids, as well as by changing the types of phospholipids used and the cholesterol content. This phenomenon appears to apply to most organisms, although it has been studied in only a few species, including bacteria, plants, and animals. Wheat grown in a cold climate will have a greater ratio of unsaturated to saturated fatty acids than the same variety of wheat grown in a warmer climate. Even within warm-blooded animals there may be differences from one tissue to another in response to temperature fluctuations. For example, reindeer were found to have a higher ratio of unsaturated to saturated fatty acids in the feet compared with the thigh muscle, where foot temperature would tend to be cooler than that of the rest of the body (Zubay 1998, 548).

A fairly new development in understanding the phospholipid membrane is the idea of lipid rafts. A lipid raft is a dynamic assembly of lipids consisting primarily of cholesterol and sphingolipids on the outer surface of the bilayer membrane of cells (Simons and Toomre 2000). It is not certain whether there is an extension of the unique physical properties of rafts to the inner layer of the membrane, but it seems likely there would be, even though the inner surface would have different components, properties, and behavior. Our understanding of lipid rafts is still in its infancy, but these unique regions appear to be playing vital roles in cell signaling, the entry of nutrients, export of products and wastes, and a wide range of other cellular functions. One can think of lipid rafts as something like peat bogs floating on a lake, where the bog is a domain with properties and features very different from those of the surrounding lake.

Cholesterol is a major component of lipid rafts, although their composition depends on how they are isolated from cells or the methods used to study them in membranes. Rafts are generally viewed as an ordered, gel-like or semisolid entity in a sea of disordered liquid or fluid phospholipids. The latter would contain an abundance of unsaturated and polyunsaturated fatty acids. Rafts tend to be composed mostly of sphingolipids with a high level of saturated or monounsaturated fatty acids and cholesterol (Simons and Toomre 2000). The cholesterol content of cell membranes varies widely, with a high of 20 to 30 percent in the plasma (outer) membrane of some cells and much lower levels (less than 10 percent) in membranes of some organelles within cells, such as nuclear, lysosomal, and endoplasmic reticulum membranes. The lipid rafts tend to be primarily in the plasma membranes of cells, where most of the cholesterol is located and where most of the signaling and processes such as endocytosis (internalizing sections of membrane that have receptors for various proteins) are taking place.

It is important to keep in mind that cells, and the body as a whole, tend to regulate cholesterol levels in response to the availability of polyunsaturated fatty acids relative to monounsaturated or saturated fatty acids. A diet rich in polyunsaturated fatty acids generally causes cholesterol levels in the bloodstream to be lower, whereas an abundance of saturated fatty acids tends to increase levels of cholesterol in the blood. The mechanisms controlling levels of cholesterol in the bloodstream are complex and only beginning to be understood. Some individuals experience large fluctuations of serum cholesterol when they change their intake of dietary fats, while cholesterol levels fluctuate very little in others who alter their diet in a similar way.

Membrane Dynamics and Eicosanoids

Cells are constantly communicating with their environment. A hormone, a drug, or a physical disturbance such as pressure or heat may cause something to happen inside the cell. When a signal such as a hormone or drug reaches a cell membrane, it is not necessary for that chemical substance to cross the membrane in order to get its signal into the cell. Instead, a series of events takes place in the lipid membrane that triggers additional events within the cell, causing changes in its metabolic state or its functional status, such as switching from idle to active. This will be discussed at greater length in chapter 4.

Cell membranes are not merely a static barrier separating the inside of a cell from its surrounding environment but are in a constant state of flux in many cells. Not only are membrane components responsible for selectively bringing necessary nutrients into cells and selectively removing waste and other products from cells, but in many cases they are constantly communicating with the environment through molecular messengers. Incoming messages are often received by protein receptors on the surface of the membranes. These protein receptors usually pass through the membranes and are anchored at a specific location in the membrane. The receptor proteins are in contact with other proteins on the inner surface of the membrane, which elicit changes dictated by the incoming signals. In many cases, the secondary protein corresponding with the receptor is an enzyme that catalyzes a chemical transformation of one or another of the phospholipids within the membrane.

An enzyme known as phospholipase A_2 (PLA_2) is one such enzyme linked to a variety of receptors. This enzyme removes one fatty acid from the middle of a phospholipid molecule. The fatty acid in question is generally a long-chain (twenty carbon atoms) polyunsaturated fatty acid, which will undergo subsequent chemical transformations through assemblies of other enzymes (see figure 2.8). As indicated earlier, one of the most common polyunsaturated fatty acids in membranes of many cells is arachidonic acid. It gets converted to a

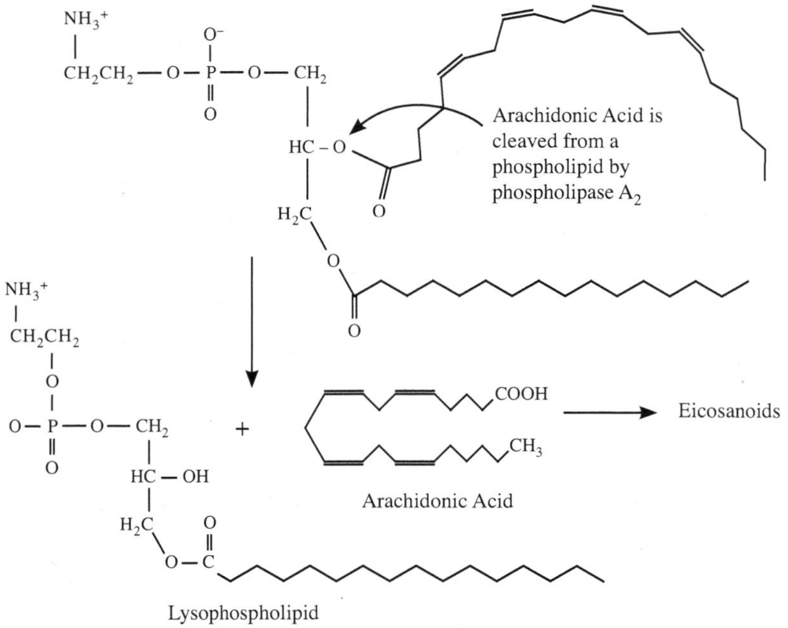

FIGURE 2.8.   The splitting of a phospholipid by phospholipase A_2, forming arachidonic acid and a lysophospholipid. The arachidonic acid can be further metabolized to bioactive eicosanoids.

wide range of extremely potent messengers in cells. Any particular cell will have a battery of enzymes responsible for transforming arachidonic acid to one or more of its many bioactive end products. (Many possible products are shown in appendix C). These end products are generally categorized according to their chemical structures and include prostaglandins, thromboxanes, leukotrienes, lipoxins, and cannabinoids.

As if this weren't complex enough, it is possible for other long-chain polyunsaturated fatty acids, such as the omega-3 eicosapentaenoic acid (EPA) and omega-6 dihomo-gamma-linolenic acid (DHGLA), to undergo analogous transformations by many of the same enzymes to form their own array of bioactive messengers. Those derived from EPA have one more double bond (see appendix C), while DHGLA products have one less double bond than those from arachidonic acid. These lead to subtle differences in the physiological activities between one series of fatty acid derivatives and another. Enzymatic transformations that require a double bond on C5 will not occur with DHGLA, because it lacks a double bond in that position.

The aggregate of all these metabolic products from the twenty-carbon polyunsaturated fatty acids are collectively known as eicosanoids (from Greek *ikosa* for twenty). They can initiate very powerful actions as signaling agents that are the foundation for numerous beneficial as well as deleterious phenomena.

Serum Lipoproteins

When blood cells are removed from the blood, the remaining fluid is known as plasma. If the blood is allowed to clot before the cells are removed, the remaining fluid is known as serum. The fluid of blood is mostly water, and since lipids are not soluble in water, they must be transported through the bloodstream in particles that remain suspended like droplets in the blood. These lipid transporting particles are known as lipoproteins, because they are composed of lipids and proteins. There are four major classes of lipoproteins: high-density lipoproteins (HDL), low-density lipoproteins (LDL), very low-density lipoproteins (VLDL), and chylomicra (or chylomicrons). Another minor lipoprotein is intermediate-density lipoprotein (IDL), but this is considered to be a remnant of VLDL and is not often discussed as a separate entity. These lipoproteins can be separated by centrifugation based on their slightly different particle densities. The sizes of the lipoprotein particles vary much more than their densities, with HDL being the smallest and chylomicra being by far the largest.

Lipoprotein particles have a monolayer (single layer) of phospholipid as a shell surrounding the particles, in contrast to a double layer of phospholipid that forms cell membranes. The interiors of lipoprotein particles are filled with lipids rather than water, so nonpolar fatty acid chains of the phospholipids mix with other lipids in the core, while the polar phosphate groups mix with water surrounding the particles. The phospholipid shell has proteins embedded that bind to receptors on the surface of cells (see figure D.1 in appendix D). Once bound to the receptors, they are brought into the cells by endocytosis, whereby the lipoprotein becomes surrounded by cell membrane and is internalized.

The embedded proteins are called apolipoproteins, meaning proteins without the lipids attached (protein only). They are classified A, B, C, D, and E, with several subgroups within some classes, such as ApoA-I, ApoA-II, ApoA-IV, ApoB-48, and ApoB-100, to name a few. Each lipoprotein particle will have more than one apolipoprotein as a constituent. The apoproteins associated with different lipoproteins are listed in appendix D. It is not necessary to understand all the different proteins involved in order to appreciate the general characteristics of these lipoproteins and how they work. An important thing to keep in mind, regarding both the apolipoproteins and the receptors that recognize and bind them, is that the information for the cell to make these proteins is carried in the genes (the DNA). There have been many different genetic variations found for apolipoproteins as well as for their receptors. Some variants result in higher levels of LDL in the blood, others in lower levels of HDL. Either type of variant would place an individual with those genes at greater risk of heart disease.

Other genetic variations have been identified that can affect lipoprotein processing and utilization. A variation in the gene for cholesteryl ester transfer protein, which is involved in the regulation of lipoprotein particle sizes, may impart exceptional longevity to those possessing this genetic variant (Barzilai et al. 2003). The main point is that genetic factors play a major role in the levels of circulating lipoproteins, their relative sizes, and their composition, all of which can have profound influences on a person's health and risk for life-threatening conditions. There will be further discussion of this in chapter 7.

Vitamin E

Vitamin E is classified as an antioxidant because it helps to protect cell membranes from spontaneous oxidation. Polyunsaturated fatty acids are very susceptible to chemical oxidation in a process that involves peroxides, hence the term lipid peroxidation (discussed in chapter 5). Since plants are the main producers of polyunsaturated fatty acids, they evolved with a system for protecting those fatty acids from lipid peroxidation. Plants that produced vitamin E (alpha-tocopherol) survived better than those that lacked this antioxidant. Vitamin E has a natural affinity for lipids and gets interspersed in membranes and fat stores. However, such a perception of random placement may not be accurate, as researchers are finding with so many biological molecules.

Since cells can regulate fluidity of their membranes by selecting saturated or unsaturated fatty acids to incorporate in the phospholipids, and also regulate the amount of cholesterol in the membranes, we may find that they also selectively distribute the available vitamin E. Vitamin E can be stored in the liver and may be mobilized from those stores in times of greater oxidative stress or need. There will be further discussion of vitamin E in chapter 5 (on oxidations), although there is still much to be learned about this important lipid antioxidant.

The Cholesterol Advantage

Cholesterol is classified as a steroid. Through the ages cholesterol evolved into an essential component of animals, being metabolized into a wide range of useful substances in the body. These include steroid hormones (progesterone, testosterone, aldosterone, cortisol, cortisone, estrogens), vitamin D (the sunshine vitamin), and bile acids or bile salts. Bile salts and phospholipids act as detergents in the digestive system, emulsifying oils and fats to make it easier for the digestive enzymes (lipases) to do their job. These polar lipids work like detergents in a washing machine, which break up the oil and grease in our clothes and get them to mix with water. Lecithin is often promoted as an aid to digestion because of this detergent action—in addition to providing choline to make the neurotransmitter acetylcholine.

Animals and humans are able to synthesize cholesterol from other nutrients, such as fatty acids (derived from triglycerides), carbohydrates, or even from protein, depending on a person's diet and the relative availability of each of these nutrient precursors. It is estimated that adults on a low-cholesterol diet synthesize as much as 1 gram of cholesterol per day to satisfy the body's needs (Glew 2002, 751). One of the actions of the lipid-lowering drugs known as statins is to inhibit an enzyme involved in the synthesis of cholesterol from other nutrients (see chapter 8).

Cholesterol is constantly being moved about as it is needed in one organ or another for many vital processes. We are gradually finding more and more that cholesterol is important for many physiological functions and consequently there is a high turnover of cholesterol each day in a normal healthy individual. Some of the systems responsible for handling cholesterol, such as the lipoproteins (especially VLDL, LDL, and HDL), may be malfunctioning for a variety of reasons. These systems may need to be brought back in line by treating the underlying cause. Dietary levels of cholesterol are rarely the underlying cause, unless very high levels are being consumed.

Our early ancestors were not concerned with their cholesterol levels, and atherosclerotic plaques were unheard of in their time. It did not matter much anyway, since the average life expectancy in those treacherous times was probably not more than forty years—hardly enough time to get a good buildup of plaque in the coronary arteries. It was not until the twentieth century that we were quite successful at pushing life expectancy to a level where many offspring of early survivors with high blood cholesterol would have sufficient time to deposit plaque in the arteries. The technological advances spawned in the medical field in the latter part of the twentieth century allow us to suddenly begin to be concerned about blood cholesterol levels, atherosclerosis, and associated health complications that threaten our longevity. However, we must keep an open mind and understand that dietary cholesterol per se is not the underlying cause of this problem, nor are saturated fats, which are so frequently implicated.

Conclusions

There are many other lipids with a host of different biological functions that are important in nature, but this chapter was intended to give a foundation for readers with little or no preparation in biochemistry (or to serve as a refresher for those who have had some exposure to the subject). Hopefully, this will provide enough background in this complex area for readers to understand the discussions that follow.

It is essential to understand the effects that saturated, monounsaturated, and polyunsaturated fatty acids have on health in order to make sound dietary choices. Triglycerides in storage fats contain these fatty acids, and they are incorporated into the phospholipids of cell membranes and lipoproteins. Cell membranes are not only a barrier between the inside and outside of a cell but also contain specialized components to transport nutrients and wastes in and out of the cells. They are involved in communication through the action of hormones, neurotransmitters, drugs, and other stimuli and are in a constant state of dynamic flux, changing all the time in response to other events taking place in the cell and in its surroundings. The membranes are also reservoirs for long-chain polyunsaturated fatty acids, which are released from the membrane and metabolized to a wide range of potent signaling agents such as bioactive eicosanoids and other lipids.

3

Some Dietary Fats Are Essential

The idea that certain foods contain substances that relieve the symptoms of what we recognize today as vitamin deficiencies dates back to ancient times. An Egyptian papyrus (Eber's Papyrus) containing a medical treatise dating to about 1500 B.C.E. recommends eating liver to cure the condition of night blindness. We know today that night blindness is caused by a deficiency of vitamin A and that liver is a good source of this vitamin. The famous Greek philosopher Hippocrates, who lived around 400 B.C.E., prescribed ox liver to cure night blindness (Liu and Roels 1980).

Eber's Papyrus also describes the disease that we now know as scurvy, although little was known about its cure at that time. Scurvy was a serious problem for naval crews during the period of great global explorations, because the sailors would be at sea, often for months, without fresh fruits or vegetables. The Chinese were well aware of scurvy, and their ships plying the seas in the late fourteenth and early fifteenth centuries grew soybeans in tubs, which provided sprouts that had ample amounts of vitamins, including vitamin C (Menzies 2003). The French explorer Jacques Cartier lost many of his crew along the St. Lawrence River in what is now Quebec in the winter of 1535–1536 when they fell ill with scurvy. The local natives brewed an extract from an evergreen (white cedar) tree that cured the survivors (Pauling 1970). The tree became known as arborvitae (Latin for tree of life) and is now a popular landscaping tree in suburban North America.

Sir Richard Hawkins, a sixteenth-century British admiral, described the use of oranges and lemons in treating British sailors who had scurvy. A later treatise by the Scottish naval surgeon James Lind, in 1753, is probably responsible for bringing this latter cure to the attention of a much wider audience, including the medical profession. British sailors became known as limeys during the nineteenth century because they were required to drink lime juice to ward off scurvy. The active component in citrus fruits was referred to as ascorbutic factor (scorbutic meaning related to scurvy and the a- prefix meaning without), and when it was finally isolated and its chemical structure determined by Albert Szent-Gyorgi in 1935, it was named ascorbic acid (Pauling 1970).

Many different organic (or plant-derived) substances were identified as being essential, although the chemical compositions of the essential components were not identified until the twentieth century. Most were chemically classified as amines because they contained the element nitrogen and were vital for human health, ergo the term vitamine. Later, when it was discovered that not all contained nitrogen, and consequently were not amines, it was perhaps deemed simpler to drop the final "e" rather than coin a new term.

In 1929 George and Mildred Burr reported that they could induce a deficiency syndrome in laboratory rats by the rigorous exclusion of fat from their diets (Burr and Burr 1929). The deficient rats showed abnormal growth and developed skin disorders, such as lesions on their tails, dandruff, loss of hair, and abnormal appearance. There were disturbances in ovulation in females and prolapsed penises in males. It was also believed, because of bloody urine, that the deficiency resulted in kidney degeneration. Adding small amounts of lard to the diet cured the deficiency syndrome, as did a small amount of liver. The researchers later tested specific fatty acids and found that adding linoleic acid or arachidonic acid (two omega-6 polyunsaturated fatty acids) to the diets would relieve the deficiency symptoms (Burr and Burr 1930). It was also found that linolenic acid, an omega-3 polyunsaturated fatty acid, could restore normal growth but did not alleviate the skin disorders. It was concluded that these three polyunsaturated fatty acids were essential in the diet and the body could not make them.

In the 1950s James Mead and coworkers at University of California, Los Angeles, worked out a metabolic pathway for the conversion of linoleic acid to arachidonic acid in humans (see figure 2.4, previous chapter). It was recognized that arachidonic acid was not essential in the diet if linoleic acid was provided (Steinberg et al. 1956). Consequently, some nutritionists chose to eliminate arachidonic acid as an essential fatty acid, leaving linoleic and linolenic acids as the only two essential fatty acids. The reasoning was that our bodies can synthesize arachidonic acid from linoleic acid but cannot carry out the reverse process. There is no good evidence that linoleic acid is needed if arachidonic acid is available, but linoleic acid is more abundant and consequently easier to obtain to satisfy our dietary essential fatty acid needs.

Similarly, linolenic acid is metabolized to eicosapentaenoic acid (EPA) and docosahexaenoic acid (DHA), which are important omega-3 fatty acids. The conversion of EPA to DHA occurs by a metabolic pathway that turned out to be different than had been expected, involving an extra elongation step and then removal of the two carbons that were just added (as shown in figure 3.1). Today it is recognized that arachidonic acid, EPA, and DHA are necessary for proper growth and development and that linoleic acid and linolenic acid can be metabolized to form these important fatty acids. Some newborn infants, however, cannot convert linolenic acid to EPA and DHA at a rate needed to satisfy normal requirements for growth and development (Heird and Lapillonne 2005). Consequently, these latter fatty acids, particularly DHA, are essential in their diet to assure proper neurological development (see chapter 11).

Essential Functions of Polyunsaturated Fatty Acids

Fatty acids are incorporated into triglycerides for fat storage in the body or into phospholipids that make up the membranes of cells. Polyunsaturated fatty acids (PUFAs) are not essential for making triglycerides, but it was discovered that

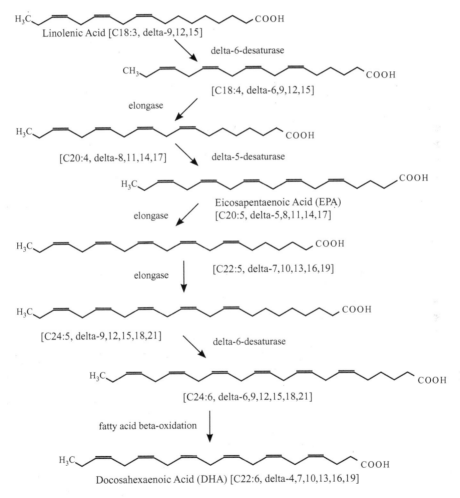

FIGURE 3.1. Metabolism of linolenic acid to docosahexaenoic acid (DHA). See text for description of this pathway.

arachidonic acid, as well as EPA and another twenty-carbon polyunsaturated fatty acid, dihomo-gamma-linolenic acid (DHGLA, an omega-6 eicosatrienoic acid), are released from the phospholipids in membranes upon specific types of stimulation. These three polyunsaturated fatty acids get metabolized to a wide array of potent bioactive substances known collectively as eicosanoids (see chapter 2). Because linoleic acid is the most abundant polyunsaturated fatty acid in terrestrial food sources, arachidonic acid is the predominant eicosanoid precursor found in terrestrial animals and humans and receives the most attention in this context. An overview of arachidonic acid metabolism to various eicosanoids with their physiological actions is illustrated in figure 3.2.

EPA and DHGLA are also metabolized to bioactive eicosanoids that are not shown in figure 3.2. EPA has one more double bond than arachidonic acid, and DHGLA has one less than arachidonic acid. Consequently, the eicosanoids formed from these fatty acids will have one more or one less double bond, which can have a profound effect on their biological activity, as will be seen later. Diagrams of the chemical structures of numerous eicosanoids produced from arachidonic acid, EPA, and DHGLA are shown in appendix C.

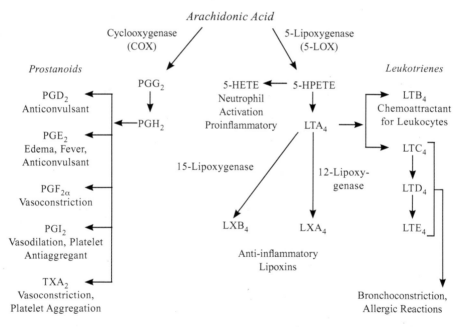

FIGURE 3.2.   Metabolism of arachidonic acid to bioactive eicosanoids. The cyclooxygenase pathway leads to prostaglandins and thromboxanes (prostanoids), and the lipoxygenase pathway leads to leukotrienes and lipoxins. Each kind of eicosanoid is formed by a different sequence of enzyme catalyzed reactions, as a result of certain cells expressing (synthesizing) specific enzymes to synthesize a particular eicosanoid or eicosanoids.

Two groups of important enzymes in eicosanoid metabolism are cyclo-oxygenases, better known as COX, and lipoxygenases (LOX). Cyclooxygenase converts arachidonic acid into prostaglandin H_2 (PGH_2), which is then converted by other enzymes into compounds known as prostanoids (prosta-glandins and thromboxanes), shown on the left side of figure 3.2. These are powerful hormone-like substances that have numerous physiological functions in the body. Most tissues have cyclooxygenase, but the enzymes for subsequent conversion of PGH_2 to other bioactive products vary from one tissue or cell type to another. For example, blood platelets have the enzymes necessary to produce thromboxane A_2 (TXA_2), which elicits platelet aggregation and vasoconstriction (narrowing of blood vessels). Vascular cells lining the blood vessels contain a different battery of enzymes that make prostacyclin (PGI_2), which impedes platelet aggregation and dilates the blood vessels. These prostanoids function in concert to maintain homeostasis in the vascular system, initiating a clot when a blood vessel is damaged but suppressing clots beyond the site of injury. Other physiological effects of the eicosanoids are shown in figure 3.2.

There are three major lipoxygenase enzymes that catalyze reactions of arachidonic acid: 5-, 12-, and 15-lipoxygenases, where the number preceding the name indicates the carbon atom of the fatty acid chain where oxygen is added. Lipoxygenases can metabolize arachidonic acid in sequence to give multiple products, as seen on the right side of figure 3.2. Metabolism of leukotriene A_4 (LTA_4) by 12-lipoxygenase produces lipoxin A_4, whereas metabolism of LTA_4 by 15-lipoxygenase gives lipoxin B_4. Both lipoxins have anti-inflammatory activity, whereas other leukotrienes derived from LTA_4 tend to be proinflammatory or involved in allergic reactions. Eicosanoids are produced in most tissues of the body, but their functions are not always clear.

The history of the discovery of prostaglandins and the derivation of this name is worth telling. In the 1930s Raphael Kurzok and Charles Lieb in the United States showed that human semen could either relax or contract strips of uterine tissue, depending on whether the tissue was from sterile or fertile women (Kurzok and Lieb 1930). Others later observed similar effects with seminal plasma (semen with the sperm cells and any other cells removed) using sheep seminal vesicles (part of the gland that secretes semen). Ulf von Euler of the Karolinska Institute in Sweden deduced that the active component was produced in the prostate gland and hence derived the name prostaglandin (von Euler 1983). Despite research on these compounds over the past seventy years and the fact that human seminal fluid has about thirty thousand times higher concentrations of E-type prostaglandins than any other body fluids, the physi-ological role for prostaglandins in the seminal fluid remains a mystery (Serhan 2001).

Later studies showed that these so-called prostaglandins are produced nearly everywhere in the body, but the name prostaglandin stuck. In the 1950s and 1960s Sune Bergström, working with von Euler, managed to isolate and chemically identify several prostaglandins and showed they were formed from arachidonic acid in active tissues. Another colleague of theirs at the Karolinska Institute, Bengt Samuelsson, studied the metabolic fate and physiological disposition of various prostaglandins and other bioactive eicosanoids derived from arachidonic acid. John Vane and colleagues at the Wellcome Research Laboratory in England discovered that anti-inflammatory agents, such as aspirin, block the production of prostaglandins and thromboxanes from arachidonic acid. The ability to block this metabolism of arachidonic acid is one of the primary biochemical actions of aspirin—indicating the wide range of effects these prostaglandins and thromboxanes can have, since this wonder drug can alleviate so many bodily discomforts. There will be more in later chapters on the events triggered by thromboxanes and prostaglandins involved in blood clotting and anticlotting.

The fact that aspirin has been used for so many maladies throughout the twentieth century would lead one to think that eicosanoids formed from arachidonic acid do mostly bad things in our bodies. However, there are many physiological actions attributed to eicosanoids, including uterine contractions or relaxations; lowering or raising blood pressure; causing platelet aggregation in the initiation of a blood clot or inhibiting platelet aggregation to prevent excessive clotting; augmenting or suppressing an immune response that may lead to inflammation; and bronchial relaxation or constriction during an asthmatic attack. Notice there are opposing effects listed in each case. The actions of the various eicosanoids are many and often diametrically opposed to one another, which has had a profound impact on making our understanding of their actions confusing, or at best complicated. These complex effects captured news headlines in 2004 when the aspirin substitutes Vioxx and Bextra (COX2 inhibitors) were pulled from the market because they could cause a heart attack rather than preventing one, as aspirin does. There will be more on this in chapter 9 (on inflammation and anti-inflammatory drugs).

The Omega Factor

Fish are cold-blooded creatures that have long been a food source for humans and other mammals. Like other living things, they have cells surrounded by phospholipid membranes, and they store excess energy as fats in the form of triglycerides. Because they are cold-blooded, they evolved with a mechanism to keep the lipids in their bodies fluid, even at the cold temperatures of the polar oceans. As discussed in the previous chapter, the more unsaturated (polyunsaturated) the fatty acids are, the lower their freezing point, or temperature where they solidify. Proper fluidity is maintained in cell membranes of fish at cold temperatures by incorporating fatty acids with more double bonds. This correlation between coldness and degree of polyunsaturation in the fatty acids is not unique to fish. Fish consume algae that grow in their environment, and algae growing in a cold environment also produce more of the polyunsaturated fatty acids that end up in fish that eat the algae, and in the fish that eat those fish.

As a general rule, the double bonds found in the fatty acids of fish extend to the third carbon from the methyl end of the chain, so they are designated omega-3 polyunsaturated fatty acids. The polyunsaturated fatty acids from terrestrial plants are predominantly ones with double bonds extending to the sixth carbon from the methyl end of the chain, and they are designated omega-6 polyunsaturated fatty acids. Oils from some plants, including soy and canola, contain some omega-3 fatty acids, and fish contain omega-6 fatty acids, but these are relatively minor (less than 10 percent) constituents of the fats from those respective sources.

The fatty acid composition of fish depends on the profile of fatty acids found in their food supply. Fish fed diets containing omega-6 vegetable oils will have an abundance of omega-6 fatty acids in their tissues relative to fish fed diets containing fish oils or cold-water algae (Ruyter et al. 2003). The vegetable oil diets also resulted in decreased levels of important long chain omega-3 polyunsaturated fatty acids, resulting in a dramatically decreased ratio of omega-3 to omega-6 fatty acids in these fish. It was also found that cardiac patients who consumed approximately 700 g (1.5 pounds) per week of salmon that had been fed diets containing either omega-3 fish oil or omega-6 vegetable oil had blood lipid and lipoprotein profiles after six weeks that reflected the salmon fillets and their respective feeds (Seierstad et al. 2005). Such results have profound implications regarding beneficial health effects of farmed fish relative to their wild counterparts.

EPA and DHA are omega-3 polyunsaturated fatty acids that are abundant in fish oils, especially cold-water fish. EPA, as a twenty-carbon polyunsaturated fatty acid, can be converted into bioactive eicosanoids. However, its conversion to DHA seems to be more critical in development of the nervous system and retina in the fetus and newborn infants. EPA and DHA were apparently not tested in early studies of essential fatty acids, so they were not designated as essential until more recently. The neurological effects arising from EPA and DHA deficiency may not be obvious in laboratory animals, since their requirements for brain development seem to be much less than those of humans. The neurological effects of these omega-3 long-chain polyunsaturated fatty acids are discussed at length in chapters 11, 12, and 13.

Clinical studies have shown that newborn infants have a limited ability to convert linoleic and linolenic acids to their biologically active products, arachidonic acid, EPA, and DHA (Lauritzen et al. 2001). One shudders to think of the consequences that nutritional guidebooks (written as late as the 1980s) had when they recommended a dietary supply of linoleic and linolenic acids for premature infants but did not mention the longer-chain polyunsaturated fatty acids (Snyderman 1980). It is expected that premature babies have even less ability to carry out the conversion to the longer-chain fatty acids. Concentrations of the long-chain polyunsaturated fatty acids are higher in the blood of newborn infants than in the mother's blood, indicating there is a selective delivery of these longer-chain fatty acids, relative to linoleic and linolenic acids, across the placenta.

After delivery, the infant can get these longer-chain acids from breast milk. Human breast milk contains about thirty times more DHA than milk from other mammals (such as cows), and the amount of DHA in the mother's milk can be increased when she consumes dietary supplements of omega-3 oils (Singh 2005). It is interesting to note that these longer-chain polyunsaturated fatty acids are carried from the mother's liver, where they can be formed from linolenic acid, to the placenta or the mammary tissue by low-density lipoproteins (LDL), the bane of the cardiovascular community.

Studies in laboratoy animals have shown that deprivation of omega-3 fatty acids in their diets can affect visual acuity and discrimination learning in newborn animals (Lauritzen et al. 2001). It has been postulated that malnutrition in pregnant women may give rise to essential fatty acid deficiencies, possibly causing permanent neurological damage to the fetus or, in milder cases, causing low birth weight and later problems in cognitive and intellectual development (see chapter 11).

Numerous studies have been performed to determine the effects of omega-3 oils on a wide range of diseases, including arthritis, coronary heart disease, cancer, inflammatory bowel disease, psoriasis, asthma, lupus erythematosis, multiple sclerosis, major depression, bipolar depression, and obesity (Simopoulos 2002b). The rationale for giving omega-3 polyunsaturated fatty acid supplements is that they compete with the omega-6 arachidonic acid for conversion to bioactive eicosanoids, and the omega-3 eicosanoids generally have weaker biological effects. Omega-3 supplements generally improve arthritis, but the effects on other disorders are less obvious or less consistent.

A major flaw in many studies is that patients are not advised to alter their diets, so consumption of a small number of omega-3 capsules while consuming a diet rich in omega-6 oils may result in little benefit—much like adding a few teaspoons of freshwater to a glass of salt water to diminish the salty taste. There would need to be a significant shift in the amount of omega-3 fatty acids relative to omega-6 fatty acids available in body tissues when signals are received for eicosanoid production in order for the biological response to be affected. It would be prudent to advise patients wishing to try omega-3 supplements for many of these disorders to try to minimize their omega-6 vegetable oil intake.

Recent studies have shown that EPA and DHA form some novel omega-3 fatty acid derivatives known as resolvins, docosatrienes, and neuroprotectins (Serhan et al. 2004). These bioactive compounds have been shown to be involved in regulating the immune system and protecting nerve tissue from oxidative damage caused by inflammation. Aspirin modifies an enzyme (COX) that forms these compounds, causing the enzyme to make a slightly altered product known as an epimer, which differs from the natural product by a change in the arrangement of atoms around one of the carbon atoms. The aspirin-triggered epimers have potent anti-inflammatory actions just like the natural (aspirin-free) derivatives of these omega-3 compounds. There will be more on this in chapter 9.

Fish Oils and Heart Disease

Fish oils, or omega-3 polyunsaturated fatty acids, are promoted for many health benefits, some of which stem from their ability to form bioactive eicosanoids or their ability to compete with arachidonic acid in being metabolized to bioactive eicosanoids. There is a simple effect of fish oils on blood lipoprotein levels that is not dependent on production of bioactive eicosanoids: they tend to lower blood cholesterol, as any polyunsaturated fatty acids will do. It is now clear that omega-3 fatty acids bind more strongly to regulatory elements that control the expression of enzymes involved in cholesterol synthesis (Schmitz and Ecker 2008). Because omega-3 fish oils tend to have a greater effect than plant oils, they are expected to have a little stronger effect in lowering blood cholesterol. Several studies have confirmed their cholesterol-lowering effect, although there is much variation in individual responses to dietary supplements (Temple 1996).

Some of the physiological effects of fish oils on blood clotting are due to the conversion of EPA to thromboxane A_3 (TXA_3) and prostaglandin I_3 (PGI_3), instead of TXA_2 and PGI_2, which are derived from arachidonic acid (Dyerberg and Bang 1980). Since EPA has one more double bond than arachidonic acid, the eicosanoid products formed from EPA will have one more double bond than those formed from arachidonic acid (see appendix C). Such a small difference in chemical structure may seem quite trivial, but it can have a profound effect on blood clotting. TXA_3 will cause platelet aggregation and vasoconstriction (constriction of the blood vessels), but it is much weaker at either of these effects than TXA_2 (the omega-6 thromboxane derived from vegetable oils). PGI_3, on the other hand, is nearly as potent as PGI_2 at blocking both platelet aggregation and vasoconstriction. The net effect of having more EPA available in the membranes, relative to arachidonic acid, is that blood will take longer to clot. This is a second, and probably much more significant, benefit of fish oil supplements with respect to heart disease—particularly for myocardial infarction.

The major benefit of fish oils with respect to heart disease is the fact that eicosanoids from omega-3 polyunsaturated fatty acids are less likely than those from omega-6 polyunsaturated fatty acids to cause premature clotting of the blood. Their ability to lower blood cholesterol is only marginally, if at all, better than that of omega-6 polyunsaturated fatty acids. Increased consumption of the typical American diet, replete with vegetable oils, has decreased the incidence of stroke from cerebral hemorrhaging in Japan but increased the incidence of coronary and cerebral infarction in that country (Okuyama, Kobayashi, and Watanabe 1996).

The French have always been an anomaly in the eyes of U.S. cardiac health-care providers and writers because their diets are so rich in saturated fats and cholesterol, yet their incidence of heart disease is about half that in the United States. The French paradox is often explained by the fact that the French typically drink wine with their meals. Some writers have postulated that alcohol is the factor, although many Americans drink beer and liquor, as well as wine, so the alcohol argument does not hold up very well. Others argue that phytochemical antioxidants in the red wine drunk by the French prevent free radical oxidation reactions that may lead to heart disease. Since it is the poly-unsaturated fatty acids that are susceptible to free radical oxidations (discussed in chapter 5), one wonders why everyone ignores this fact when explaining the French paradox. In other words, the French are consuming lower amounts of the omega-6 polyunsaturated fatty acids than Americans, which accounts for lower amounts of polyunsaturated fatty acids available for oxidation in lipoproteins and consequently less formation of atherosclerosis (see chapter 6).

Unfortunately, much of the U.S. cardiac health community has been indifferent to the proxidant effects and eicosanoid connections regarding poly-unsaturated fatty acids. This arises because eicosanoid actions are very complex, often contradictory, and not well understood by most people outside the eico-sanoid research community. On the other hand, the cholesterol hypothesis seems to be quite easy to understand: (1) saturated fats tend to increase blood cholesterol; (2) polyunsaturated oils tend to lower blood cholesterol; (3) people with high blood cholesterol, especially LDL-cholesterol, are more likely to have a heart attack than people with low blood cholesterol. This seems as simple as one-two-three, but if one wants to understand what causes heart disease, it will be necessary to probe deeper into the factors involved.

Conclusions

Inflammation is a normal adjunct of an immune response that involves a concerted series of physiological events. The immune cells mounting an attack on invading or injured cells release a host of messengers that trigger vasodilation and increased vascular permeability, which allows leukocytes to get from the bloodstream to the affected tissue. Edema or swelling ensues from this activity, and proliferation of cells at the site, along with high metabolic activity of the leukocytes, generates heat. These effects are augmented by a signaling protein known as pyrogen. All these events involve eicosanoids at one level or another. The eicosanoids are often prerequisite, or acting upstream in the physiological sense, for the actions of cytokines and other cell-signaling mediators. There will be further discussion of many of these actions in later chapters.

Polyunsaturated fatty acids are now known to be involved in a wide range of biochemical and physiological phenomena related to many diseases, but they are often overlooked or not emphasized by professionals who are giving advice on diet and health. Genetic variation has also received wide recognition as a factor that affects blood lipids and risk for many diseases. The following chapters will bring to light many of the beneficial as well as deleterious effects of dietary polyunsaturated oils.

4

Signals, Messengers, and Responses

There are countless signals impinging on cells all the time in order to keep them in synchrony with the rest of the body. They also respond to various changes in environment and nutritional status. Whether a cell responds to a particular signal or not depends on whether that cell has the equipment to respond. Cellular equipment refers to proteins, which are responsible for most of what happens in a cell.

An analogy may help to give a clearer picture. For example, a radio station sends its signals across the airwaves, but only those who happen to have their radios turned on and tuned to the proper frequency will receive the signal for that particular station and respond, whether the response is just the pleasure of listening or being cajoled into calling the station to win a prize. Similarly, cells will respond to specific signals if they have receptors (proteins) to bind specific chemical signals (such as hormones) that may be sent in their direction. As different listeners to the radio may react to what is being sent across the airwaves in different ways, cells in the body may react to a particular hormone in different ways, depending on the type of cell (for example liver, muscle, heart, or brain).

Signals in our modern culture come in many different forms, such as radio, television, cellular phones, or remote control devices, each of which may have electromagnetic waves emitted from a source and captured by a receiver. Signals in the body also come in different forms but generally depend on chemicals secreted by one cell and received by another cell. Chemical signals secreted by specialized cells known as glands travel through the bloodstream as hormones to affect other cells throughout the body. This would be analogous to radio waves that cover a large geographical area and are available to a wide range of people.

On the other hand, neurotransmitters are typically released by one cell and travel only to an adjacent cell to accomplish their mission. This would be like a remote control device for controlling a television set that works only in close

44

proximity to the set. There are many other messengers that fall into categories between these extremes, often affecting cells in a localized area, such as the control of smooth-muscle cells in the digestive and circulatory systems or those causing pain and inflammation where there is an infection or injury. Such paracrine messengers target only specific types of cells in a localized region, which might be analogous to police band or fire department radios that cover a limited range and are available to a limited number of people.

Electrical signals in the body are extremely fast and are elicited by the movement of ions (electrolytes) across cell membranes, some of which result in a discharge of the membranes that can cause the far end of the cell to react within milliseconds. The ions involved in electrical signals are usually sodium (Na^+), potassium (K^+), calcium (Ca^{2+}), and chloride (Cl^-). These ions behave much as electricity does in wires, causing things to happen quickly. For example, when a nerve cell receives signals to communicate (send its own signal), sodium and calcium ions surge into the nerve cell at specific locations known as ion channels, ultimately causing that nerve cell to release its chemical transmitters to the next cells. Potassium and chloride ions flow across membranes through their own ion channels, generally to counteract the effects of sodium and calcium, keeping the system more or less in balance. The cell must be well charged for these ionic signals to work properly, and there are pumps in the cell to move these ions back to the compartments where they are supposed to be in order to recharge the cell—much like recharging a battery. Many of the events involved in neurotransmission (nerve impulses) are described in appendix E.

Molecular signals are generally slower than electrical signals, operating in seconds or minutes rather than in milliseconds. Many of the intercellular messengers need not enter the cells but instead bind to receptors on the surface of the cells, thereby causing things to happen inside the cell. This is like a messenger going to the entry of a high-security building. The messenger is not allowed in but gives the message to the guard at the door. The message may have to be transferred to another messenger inside the high-security area, and it may get delivered to several different recipients. An interesting system in the body that is well understood in biochemical terms is the signal sent to the liver and muscles from the adrenal gland when there is danger about. This messenger system will serve well to illustrate the general principle.

Epinephrine Receptors Exemplify Signaling Mechanisms

When we see or hear something that frightens us, the brain sends out an alarm. The adrenal gland receives signals from the brain to release epinephrine (also called adrenaline). This hormone travels through the blood and binds to specific receptors on cells that recognize this hormone. When the epinephrine binds to its receptors on the surface of liver or muscle cells, it causes an enzyme inside the cell membrane to become activated and catalyze a chemical reaction that produces another messenger known as cyclic-adenosine monophosphate (or cyclic AMP) inside the cell. Each signal burst may produce thousands of cyclic AMP molecules. Such intracellular messengers formed in response to an external signal are called second messengers.

Cyclic AMP is formed from the high-energy compound known as ATP (adenosine triphosphate). After many molecules of this second messenger are formed, these cyclic AMP molecules will activate specific proteins in the cell. One protein that cyclic AMP activates is an enzyme known as protein kinase A. This enzyme will transfer a phosphate group from ATP to several different proteins in the cell, such as enzymes, transporters, receptors, and ion channels, depending on the type of cell and its functions. Some of these proteins become more active as a result of the phosphate being attached, while others become less active, resulting in some metabolic pathways or cell functions being turned on and other metabolic pathways or cell functions being turned off.

When epinephrine binds to beta-receptors on the surface of liver cells, the levels of cyclic AMP in liver cells rises, and specific proteins become activated or inactivated (see figure 4.1). As a consequence, the metabolic pathway for storing glucose as glycogen (starch) in those cells is turned off, and the pathway for breaking down glycogen in the cell to form glucose is turned on. Glycolysis, the pathway for metabolizing glucose for energy, is turned off as well, so the liver cells will not use up the glucose that is needed for other cells in this alarm response. Epinephrine is signaling the body to prepare for fight or flight; in either case, more glucose will be needed by the muscles and heart, so the liver cooperates by releasing glucose it has stored as glycogen.

The cyclic AMP and activated protein kinase A in the heart and skeletal muscles cause enzymes for glycolysis to be turned on (the opposite of cyclic AMP's effect in the liver), resulting in activation of this pathway of glucose metabolism for energy production. The energy will be needed to get the heart pumping faster and the muscles ready to spring into action. Although glycolysis is affected in different ways in the liver (inhibited) compared with the heart and muscles (activated), the pathway for breaking down glycogen is activated in all three tissues. The heart rate (beats per minute) and force of contraction increase as a result of protein kinase A attaching phosphate groups to ion channels, decreasing the movement of potassium and enhancing calcium movement in response to sympathetic nerve stimulation.

The increased levels of cyclic AMP in some cells may ultimately have effects in the nucleus, causing certain genes to be expressed and thereby producing specific proteins that may be needed for a response. This may be how stress, which prolongs elevated levels of epinephrine in the blood, causes certain changes to take place in the body that can be longer lasting than just the activation of certain proteins for a few minutes during the fight-or-flight response. There may be many signals involved in response to stress that reinforce one another or cancel other signals.

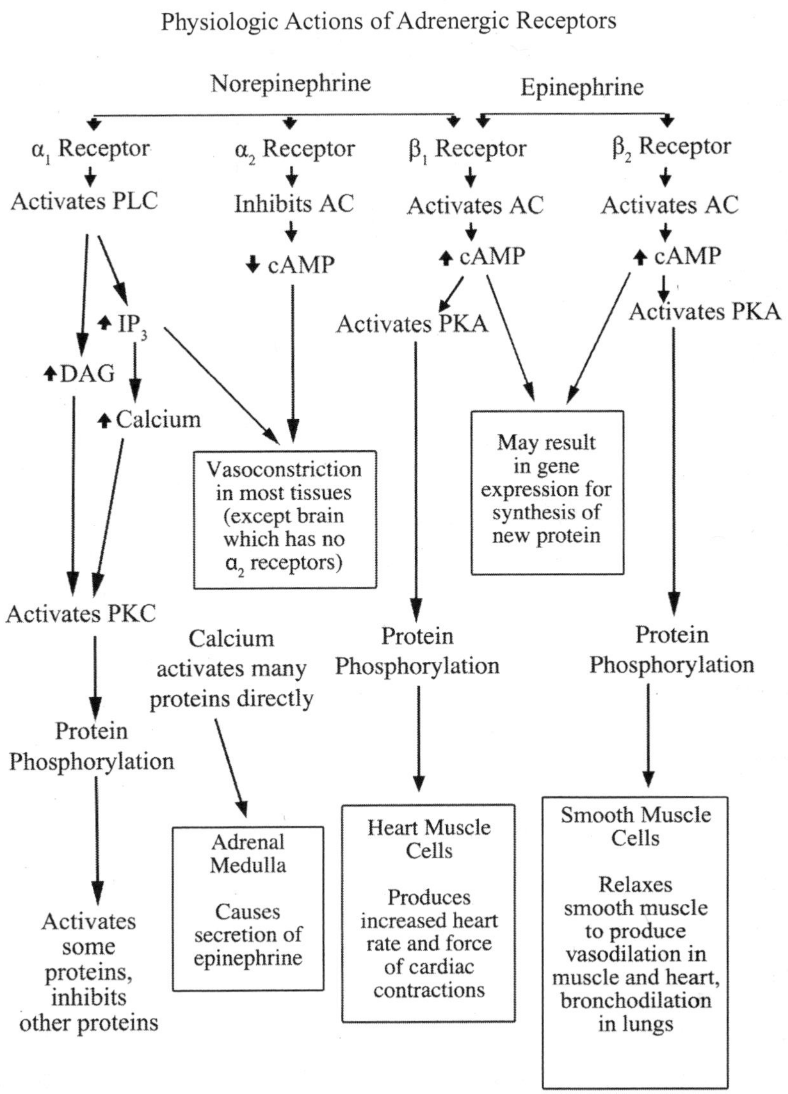

Physiologic Actions of Adrenergic Receptors

Norepinephrine   Epinephrine

α_1 Receptor α_2 Receptor β_1 Receptor β_2 Receptor

Activates PLC   Inhibits AC   Activates AC   Activates AC

$\downarrow$ cAMP $\uparrow$ cAMP $\uparrow$ cAMP

$\uparrow$ IP$_3$ Activates PKA   Activates PKA

$\uparrow$ DAG

$\uparrow$ Calcium

Vasoconstriction in most tissues (except brain which has no α_2 receptors)

May result in gene expression for synthesis of new protein

Activates PKC

Calcium activates many proteins directly   Protein Phosphorylation   Protein Phosphorylation

Protein Phosphorylation

Activates some proteins, inhibits other proteins

Adrenal Medulla

Causes secretion of epinephrine

Heart Muscle Cells

Produces increased heart rate and force of cardiac contractions

Smooth Muscle Cells

Relaxes smooth muscle to produce vasodilation in muscle and heart, bronchodilation in lungs

FIGURE 4.1. The four major receptors for epinephrine and norepinephrine. The physiological actions in specific tissues and resulting biological responses are shown.

Other cells in the body may respond in a completely different manner compared with liver, muscle, or heart cells. For example, some cells in other tissues make receptors for epinephrine that inhibit the enzyme that makes cyclic AMP (see the alpha-2 receptors in figure 4.1). This would result in blocking the events described above for the beta-receptors. Furthermore, epinephrine is not the only hormone that activates the enzyme that forms cyclic AMP, so what seems like a contradictory action of epinephrine in some cells may be a control to keep the response of certain cells in check when epinephrine has been secreted into the blood. Those cells might be responding to some other signal at the time the epinephrine signal is received, so epinephrine may modulate the other signal, making it stronger or weaker. Some blood vessels tend to constrict, while blood vessels in the heart and skeletal muscles dilate. Airways in the lungs dilate to reduce resistance to airflow during breathing.

The alpha-1 receptors are linked to a different enzyme known as phospholipase C, which splits the phospholipid phosphatidyl inositol 4,5-bisphosphate (PIP_3) into two pieces, one being inositol triphosphate (IP_3) and the other diacylglycerol (DAG) (see figure 4.2). Inositol triphosphate acts as a second messenger within the cell to release calcium from intracellular storage sites. The released calcium causes a wide range of changes to take place inside the cell, by affecting the activity of specific proteins. Diacylglycerol cleaved from the other half of the phospholipid activates an enzyme known as protein kinase C, which attaches phosphate to a wide range of proteins in the cell, causing those proteins to become activated or inactivated. Protein kinase C and protein kinase A (activated by cyclic AMP, as described above for the beta-receptors) both attach phosphates to many different proteins in cells. The subsequent effects depend on the specific proteins and their functions in the overall scheme of metabolism and communication. Most proteins phosphorylated by protein kinase A will not be phosphorylated by protein kinase C and vice versa, but some proteins may be phosphorylated by both. The activation of phospholipase C by an incoming signal sparks a two-pronged response in the receiving cell—one because of a sudden increase in calcium levels in the cell, the other because of activation of protein kinase C and attachment of phosphate to numerous proteins, altering their activities or functions.

Despite the inherent complexity of such disparate and conflicting actions related to the epinephrine system, they make sense in the context of the bigger picture of a body fine-tuning its responses to the environment. At least four different types of receptors have been characterized for epinephrine: alpha-1 (α_1), alpha-2 (α_2), beta-1 (β_1), and beta-2 (β_2). There are others, but they are less abundant and less understood. Each of these four also work as receptors for the sympathetic neurotransmitter norepinephrine (also called noradrenaline), although some favor epinephrine while others prefer norepinephrine in terms of their binding affinity and distribution in the body. Epinephrine is primarily a hormone that circulates throughout the body, whereas norepinephrine is primarily a neurotransmitter that sends signals across a synapse to an adjacent cell. The sympathetic nervous system stimulates the adrenal medulla to secrete epinephrine, so these similar catecholamines work together in a kind of synergy. Epinephrine and norepinephrine have similar chemical structures, so it is not surprising that they bind to the same receptors.

There are a phenomenal number of messengers in the body, with ten or so of the classical amine and amino acid neurotransmitters and hormones (such as

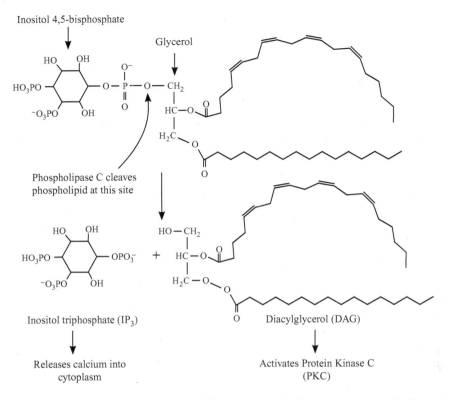

FIGURE 4.2. Splitting of phosphatidyl inositol-4,5-bisphosphate by phospholipase C to produce inositol triphosphate (IP3) and diacylglycerol (DAG).

epinephrine, norepinephrine, dopamine, serotonin, acetylcholine, histamine, glutamate, and gamma-aminobutyric acid or GABA). There are nearly one hundred different peptide neurotransmitters and hormones that have been identified (insulin, glucagon, endorphins, and adrenocorticotrophic hormone or ACTH, to name only a few). Several of these peptide hormones are formed from the same original protein (such as beta-endorphin and ACTH) but differ slightly in the way the protein is processed from one type of cell to another. There are about ten different steroid hormones formed from cholesterol, and scientists are finding more and more messengers that are being formed from polyunsaturated fatty acids. Since this book focuses on lipids, some of these lipid messengers will be discussed in the context of how they work.

Lipid Messengers

Lipid messengers differ from water-soluble messengers in that they can, and usually do, cross cell membranes and bind to receptors either inside the receiving cell or possibly on the surface of the cell. For example, some steroid hormones traveling through the bloodstream can diffuse through the plasma membrane of a cell and eventually penetrate the nuclear membrane. Steroid receptors are usually located in the nucleus of the cell, where the receptor binds the steroid hormone and the steroid hormone-receptor complex then binds to a specific site on DNA known as the hormone response element, as shown in figure 4.3. This allows a segment of the DNA (a gene) to be expressed, resulting in synthesis of a specific protein in the cell. An unoccupied steroid receptor may bind to the DNA and turn off the expression of certain genes. When the steroid arrives, it releases the lock on those genes. This is how many steroid hormones and related steroid drugs work.

Testosterone, the male steroid sex hormone, for example, can bind to receptors in muscle cells, allowing the genes for muscle proteins to be expressed and thereby making more muscle mass, which is characteristic of males. Other signals, such as weight-bearing exercise, must also occur for optimum effect. Many athletes have found that anabolic steroids (synthetic substances that mimic the actions of testosterone) will help them bulk up in muscle mass more than they would from the natural levels of testosterone that are present. Because women produce much less testosterone than men, their bodies tend not to produce as much muscle mass, even when they do strenuous exercise. However, anabolic steroids will cause women to build more muscle mass with strenuous exercise, but they may also induce many undesirable effects that go along with masculinity, such as appearance of facial hair and disruption of the menstrual cycle. Testosterone (and anabolic steroids) can have many effects

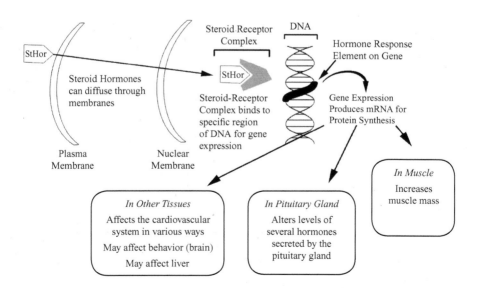

FIGURE 4.3. The action of steroid hormones on their nuclear receptors to cause gene expression and synthesis of new protein.

in different types of cells in the body, in addition to their anabolic (tissue-building) effects in muscle.

There is a family of proteins that bind cholesterol and regulate expression of multiple genes for proteins or enzymes involved in cholesterol biosynthesis and transport. They were given the descriptive name of sterol regulatory element binding proteins (SREBP) and belong to a family known as leucine zipper transcription factors. When cholesterol levels are high in the cell, cholesterol binds to SREBP-1c, and the complex moves into the nucleus to alter transcription of several genes (Brown and Goldstein 1997). This results in an increase in enzymes involved in the synthesis of fatty acids and other lipids. Long-chain (twenty or twenty-two carbons) polyunsaturated fatty acids are known to suppress expression of SREBP-1c, resulting in decreased synthesis of new lipids. The long-chain polyunsaturated fatty acids bind to a variety of other nuclear transcription factors and lipid receptors to alter a wide range of gene expression (Jump 2008). In this way dietary polyunsaturated oils alter not only the profile of blood lipids (cholesterol, LDL, VLDL, and HDL) but also affect several cellular systems.

There are many other lipid messengers that have a wide range of effects on different cells, many of which are not completely understood in terms of their biochemical processes. Because these messengers can diffuse across cell membranes, they are able to enter cells in the region where they are produced. They may be produced in the nervous system in response to neurotransmission, where a neurotransmitter triggers a cascade of second messengers in the receiving cell that subsequently activate enzymes to form a lipid messenger. The lipid messenger may then drift back to the cell that sent the original signal, resulting in some modulation or down-regulation of the signals being sent by that cell. This is known as a retrograde signal and helps to keep the nervous system in balance.

The lipid messengers tend to be extremely potent and are proving to be extremely complex in their actions as well. It was originally thought that any actions of the polyunsaturated fatty acids (PUFAs) on cells would be due to the PUFAs being metabolized by enzymes to other bioactive lipid messengers. However, there is a large body of evidence accumulating that indicates PUFAs may also be acting as messengers or signals without being metabolized (Briscoe et al. 2003; Brown, Jupe, and Briscoe 2005). Many of the effects observed regarding PUFAs may be due to their ability to get proteins to interact better with membranes and altering the activity of those proteins. Biochemists often find it difficult to tease out the mechanisms by which these fatty acids alter cellular activities because there are so many possibilities for them to get metabolized by many different enzymes. When the integrity of a cell is destroyed in order to find these molecules, the lipid messengers may become transformed as a result of the assault.

The metabolism of PUFAs to form known bioactive messengers gives rise to a staggering array of possibilities, as described in chapter 2. For example, arachidonic acid is normally found attached to the phospholipids of cell membranes. Certain signals can activate the enzyme known as phospholipase A_2 (PLA$_2$) that releases the arachidonic acid from the cell membrane, making it available to be metabolized by a variety of different enzymes in the cell where it originated or in neighboring cells where it may happen to wander. One of these enzymes is cyclooxygenase (COX), which sends the arachidonic acid along a metabolic pathway leading to either prostaglandins or thromboxanes, depending on whether the cell where this metabolism takes place contains the battery of enzymes for the former or the latter.

Blood platelets contain enzymes to convert arachidonic acid into thromboxane A_2 (TXA$_2$), which causes the platelets to stick together (aggregate) and smooth muscles surrounding the blood vessels to contract and constrict the vessel. The net result of these actions of TXA$_2$ is to initiate a blood clot. There is an onslaught of molecular events triggered by TXA$_2$ that are involved in the clotting process (see figure 4.4). These include activation of the enzyme thrombin to transform fibrinogen circulating in the bloodstream into active fibrin, which creates the protein meshwork for a blood clot. Aspirin inhibits formation of TXA$_2$ to prevent initiation of this process. Coumadin (warfarin) inhibits the transformation of fibrinogen to fibrin to block the ultimate formation of a clot after the TXA$_2$ has initiated events. Although low-dose aspirin is effective at suppressing the initiation of a blood clot, there has been a trend toward using coumadin for people at risk of premature clotting. Coumadin, originally marketed as rat poison, can prevent clot formation when there is minor internal bleeding. The physician must monitor the patient on Coumadin closely to assure the proper dosage of this dangerous drug. Careful monitoring is needed to prevent the patient from succumbing to the action of this drug that kills rats—death from internal hemorrhaging.

Vascular endothelial cells (the cells that form the blood vessels) have a different battery of enzymes that result in the arachidonic acid being converted into prostaglandin I_2 (PGI$_2$ or prostacyclin), which prevents the platelets from sticking together and relaxes the smooth muscles surrounding the blood vessels. It counteracts the effects of TXA$_2$. This may sound counterproductive, but it works well to keep the vascular system in balance. Thrombosis occurs where the vessel is leaking but not in the regions beyond. Since aspirin blocks formation of both TXA$_2$ and PGI$_2$, both the prothrombotic and antithrombotic eicosanoids are suppressed. It would seem much more reasonable to use aspirin for premature clotting problems than Coumadin, in view of the much narrower window for effective dose in the case of Coumadin. In addition, aspirin therapy is only a small fraction of the cost of Coumadin therapy and monitoring.

Arachidonic acid can be metabolized by another enzyme known as lipoxygenase, found in many kinds of white blood cells and other types of cells, sending it down a different metabolic pathway to form a family of potent messengers known as leukotrienes. These include leukotrienes A_4, B_4, and so on

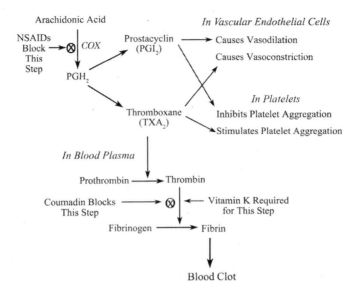

FIGURE 4.4.   The role of thromboxane A2 and prostacyclin in vascular homeostasis and thrombosis. The sites where aspirin (and other nonsteroidal anti-inflammatory drugs) prevent blood clotting and where Coumadin inhibits the conversion of fibrinogen to fibrin by blocking the action of vitamin K are shown with a $\otimes$.

to E_4. Leukotriene B_4 (LTB_4) is known as a chemotactic agent that attracts white blood cells to the site where it is produced. It is produced by cells where there has been an injury or infection. For example, if you get a splinter in your finger, the tissue surrounding the splinter produces LTB_4, which diffuses away from the site. When the LTB_4 comes in contact with a leukocyte (white blood cell) in the bloodstream, the leukocyte will squeeze through the cells lining the blood vessels and move toward the site where the LTB_4 is being produced (toward higher concentrations of LTB_4).

Other leukotrienes formed from arachidonic acid include LTC_4, LTD_4, and LTE_4, which together are known as slow-reacting substance of anaphylaxis. These are substances produced in an allergic reaction that cause alveoli of the lungs to contract, making it extremely difficult to breathe or get air into the lungs. They are responsible for asthmatic attacks. Such a life-threatening response mechanism most likely evolved to minimize the amount of noxious substances taken into the lungs when they are in the air we are breathing. This would alert us to a toxic environment and get us to move as quickly as possible to fresh air. I remember experiencing a most unpleasant sensation when lying down in a shallow pool of water at a hot spring. The sulfurous gases in the hot spring, which were hanging low above the water because they are more dense than air, triggered the production of these leukotrienes in my lungs and caused constriction in my airways. I immediately realized that it was not good for me to stay there breathing these noxious fumes and quickly got out to fresh air, where I recovered my normal breathing after a few minutes. Unfortunately, most people with severe cases of asthma are unable to get relief so easily.

The large increases in severe asthma in industrialized countries may be due to higher levels of omega-6 polyunsaturated fatty acids in the diet that become available for leukotriene production in the lungs. This will be discussed at greater length in chapter 9.

There are at least three different forms of lipoxygenase enzymes in different cells of the body, with one of them initiating the leukotriene pathway. Other bioactive products formed by these enzymes are known as HETE, short for hydroxyeicosatetraenes. These are also potent messengers that are not that well understood, but some appear to have anti-inflammatory properties, while others are proinflammatory. The metabolism of these substances may take place in more than one cell, a process known as transcellular metabolism; after one hydroxy group is placed on the arachidonic acid in one cell, it diffuses to another cell, where one or two more hydroxy groups may get attached (Serhan and Oliw 2001). There will be more about these in chapter 9.

Endocannabinoids—the Body's Own Intoxicating Lipid Messengers

Perhaps some of the more interesting metabolites of arachidonic acid are the endocannabinoids. The name is derived from the fact that they are substances produced in the body (endo) that bind to the same receptors in the body as tetrahydrocannabinol (THC), the active component of marijuana (Cannabis sativa). The first endocannabinoid to be discovered was anandamide, whose chemical name is N-arachidonylethanolamide. This lipid messenger has arachidonic acid attached to the nitrogen atom of ethanolamine. Ethanolamine is formed by decarboxylation of the amino acid serine. It is related to choline and is found in the phospholipids of membranes (phosphatidyl ethanolamine). A series of enzyme-catalyzed reactions attach three methyl groups to the nitrogen atom of ethanolamine to produce choline. Choline is not only a component of lecithin, a cell membrane phospholipid, but is also a component of the neurotransmitter acetylcholine. So both ethanolamine and choline are components of phospholipids, and each is a component of a different messenger or signaling agent.

The psychoactive effect of marijuana has been known for thousands of years, although the active component, THC, was isolated from the plant only in 1964 (Gaoni and Mechoulam 1964). The endogenous compound was not discovered until 1992 by William Devane and colleagues (Devane et al. 1992). The name anandamide was derived from the Sanskrit word *ananda*, meaning bliss, and the fact that the compound is in the chemical class known as amides. Additional compounds in the endocannabinoid family include several ethanolamine derivatives of other polyunsaturated fatty acids that appear to require at least three double bonds and at least twenty carbon atoms in the fatty acid (Mechoulam, Hanus, and Martin 1994). In addition, arachidonic acid may form an ester link to ethanolamine, meaning that the arachidonic acid gets attached to the opposite end of the ethanolamine molecule (figure 4.5). Some endocannabinoids have antagonist activity, meaning that they bind to the cannabinoid receptor but rather than activate the receptor they block it (Porter et al. 2002).

Another endogenous ligand for the cannabinoid receptor is 2-arachidonyl-glycerol and its ether analog, where arachidonic acid is attached to glycerol, a component of triglycerides and most phospholipids. The early pioneers in the cannabinoid field, Raphael Mechoulam and his colleagues at Hebrew University in Jerusalem, as well as many other investigators, have identified a wide range of compounds produced in the body that bind to cannabinoid receptors.

Cannabinoids, whether formed in the body or obtained from external sources such as marijuana or THC in pharmaceutical form, have many and varied effects in the body. They were found to relax the smooth muscles surrounding blood vessels (dilating the blood vessels) to varying degrees in different vascular preparations, either from different tissues of a species or from the same tissue in different species, and may have no effect in some vascular preparations (Randall, Kendall, and O'Sullivan 2004). It would seem that THC dilates the blood vessels in the eye, causing the bloodshot eyes or redness often seen in some individuals who smoke marijuana. However, not all individuals get bloodshot eyes when smoking marijuana.

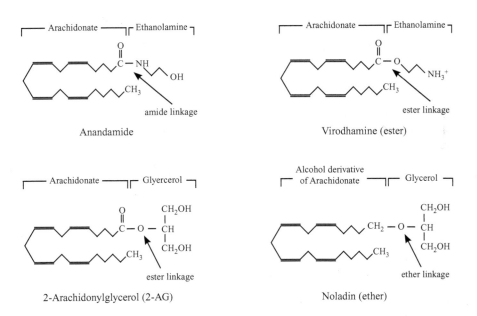

FIGURE 4.5. Structures of some endocannabinoids.

Studies on cardiovascular effects of cannabinoids have reported both increases and decreases in blood pressure. Cannabis (marijuana) can occasionally cause pronounced increases in heart rate (tachycardia) in humans, although the effect in animals is usually to slow heart rate (bradycardia). Studies in anesthetized animals are sometimes in conflict with studies in unanesthetized animals. Some studies found that cannabinoids block release of norepinephrine from neurons. Norepinephrine helps to regulate blood pressure and heart rate in complex ways.

Cannabinoids, particularly those taken from external sources, have complex actions in the brain, causing a wide range of effects on memory, cognition, spatial perception, and coordination. The cannabinoid receptor is one of the most abundant modulatory (not causing direct excitation or inhibition of neuron firing) receptors in the brain. When neurons are firing, it seems the receiving cells may produce the endocannabinoids in response to overstimulation. The endocannabinoids then diffuse back to the nerve terminals where the excitatory (or perhaps inhibitory) signal was originally sent. When the endocannabinoids bind to the CB1 cannabinoid receptor in the brain, it inhibits the production of the second messenger cyclic AMP in the nerve terminal (similar to the action of epinephrine at alpha-2 receptors discussed above). This will affect the movement of ions in and out of the nerve terminal, resulting in less release of neurotransmitter (Fride 2002). The neurons probably produce the endocannabinoids only when there is too much activity, trying to restore a balance to brain activity.

Muscular coordination can be impaired by smoking marijuana because neurons that fine-tune the signals involved in fine-motor coordination may also be hampered by THC. THC in pharmaceutical form (Marinol) has been approved by the FDA for treatment of anorexia associated with AIDS, and CB1 receptor antagonists have been tested for treatment of obesity. Although subjects taking the receptor antagonist rimonabant experienced significantly greater weight loss than subjects taking a placebo, many more drug-treated patients than placebo takers dropped out of the study because of side effects (Van Gaal et al. 2008). Side effects included nausea, depression, anxiety, and irritability (which seem to be similar to some withdrawal effects from marijuana). Clearly such drugs are not going to be the magic bullet for combating obesity.

When a person smokes marijuana, THC floods the brain, binding to the cannabinoid receptors and slowing down neural transmissions universally. This affects systems responsible for short-term memory, and when short-term memory is interrupted, long-term memory formed from the short-term memory will also be impaired. This begs the question of why we would have an endogenous system that would interrupt short-term memory. But think about all those sensory inputs entering the brain every day that you would not want cluttering your precious memory capacity. Why are some experiences selected for memory storage while others are not? There is still much to be learned about this system.

These are just some of the diverse effects that marijuana can have, but it is easy to see that one of its overall effects is to slow down neurotransmission.

Cannabinoid receptors are widespread in the body; they are found not only in the nervous system but also in spleen and lymph nodes (immune system) (Lynn and Herkenham 1994). They are abundant in areas of the brain that are responsible for thought processes (cerebral cortex), memory (hippocampus), and emotion (amygdala), although there seem to be relatively few cannabinoid receptors in the brain stem, which regulates basic life functions such as heartbeat and diaphragm movement. This may be the reason there is no evidence of anyone dying from an overdose of THC (Pollan 2001). The list of effects associated with the cannabinoid system in the body continues to grow. Studies indicate that increased activity of the cannabinoid system in the brain may be involved in the pathogenesis of schizophrenia (Ujike and Morita 2004). Smoking marijuana can produce paranoia. It has also been found that blocking or inactivating the CB1 receptors in rodents' brains can produce anxiety-like responses and aggressive behavior. Aggressive behavior may also be a withdrawal effect in some (not all) humans who smoke marijuana.

In addition to the wide range of endogenous cannabinoids produced in the body and the diversity and distribution of the cannabinoid receptors, there is also a great deal of interest in the activities of the enzymes that produce endocannabinoids, as well as those responsible for their metabolic inactivation or elimination. Genetic variations in these enzymes may lead to a wide range of physiological and psychological effects of both endogenous and extraneous cannabinoids in different individuals. Development of drugs that inhibit the enzymes would have great potential for a wide range of therapeutic effects, albeit at the expense of great potential for undesirable side effects in view of the broad distribution and complex actions of the endocannabinoids.

Nitric Oxide—Another Unconventional Messenger

The cannabinoids also seem to stimulate production of nitric oxide (NO), which until the 1980s was thought to be just a noxious pollutant in the air we breathe (Deutsch et al. 1997). It turns out to be another messenger in the body that causes relaxation of smooth muscles surrounding blood vessels and lowers blood pressure. Drugs such as nitroglycerin and amyl nitrate are often given to patients with acute angina (chest pains near the heart), which is caused by constriction of the blood vessels around the heart and can lead to a heart attack. These drugs get metabolized in the body to nitric oxide, resulting in dilation of the blood vessels and relief of the pain of angina.

Nitric oxide was also found to be a messenger that causes relaxation of the arterioles (blood vessels) in the vascular chambers in the penis, giving rise to an erection. Other factors or messengers, including the parasympathetic nervous system, are involved in the erectile response, so heart patients taking nitroglycerin or amyl nitrate for a sudden attack of angina will not necessarily get an erection. Amyl nitrate is often put in a small capsule that is popped open, and the volatile drug shoots out as a gas that can be inhaled. This form of the medication became popular, especially in the gay community, as a street drug known as "poppers" for enhanced penile erections. There is anecdotal evidence that cannabinoids or smoking marijuana can enhance male performance, perhaps through its stimulation of nitric oxide production, although scientific studies of this aspect of cannabinoid pharmacology are lacking. In view of the many and varied responses of cannabinoids, smoking marijuana would probably not give a consistent effect in this context. This is mainly due to the fact that the endocannabinoids seem to be adjunct or cosignaling agents and many other messengers are interacting with these to produce an overall neurophysiological effect.

How is it that Viagra, the drug that is widely used for erectile dysfunction or impotence, works to accomplish the same feat? The NO signal results in production of cyclic guanosine monophosphate (cyclic GMP) in the arteriolar smooth-muscle cells of the penis. Cyclic GMP is a close relative of cyclic AMP mentioned earlier. Cyclic GMP is the intracellular second messenger that brings about relaxation of the smooth muscle by altering ion channels in these cells. An enzyme known as phosphodiesterase breaks down the cyclic GMP, resulting in the ion channels returning to their previous state and contraction of the smooth muscles to constrict the arterioles. The penis then returns to its normal flaccid state. Viagra (generically known as sildenafil) inhibits phosphodiesterase, preventing this enzyme from breaking down cyclic GMP. The smooth muscles remain relaxed for a longer time, and the arterioles remain dilated to keep the vascular chambers filled with blood. Viagra has its effect downstream from nitric oxide and the production of cyclic GMP, so it will not initiate an erection but only prolongs an existing one. To borrow an analogy from Lauralee Sherwood, author of a textbook on human physiology, "Just as pushing a pedal on a piano will not cause a note to be played but will prolong a played note, sildenafil cannot cause the release of nitric oxide and subsequent activation of the erection-producing cGMP, but it can prolong the triggered response" (Sherwood 2004, 767).

Conclusions

There are a host of messengers in the body that fall into the class of peptide (small protein) neurotransmitters, hormones, and cytokines. These peptide messengers will be discussed in the context of their interactions with the lipid messengers on many occasions throughout this book. Their actions are elicited in the same way as those of other messengers, through binding to their receptor proteins, which trigger some change in enzyme activity or other alterations in the actions of proteins.

The three different modes of signal transduction represented by adrenergic (epinephrine and norepinephrine) receptors are common among the numerous hormones and neurotransmitters that will be discussed in later chapters. The action of steroids on their nuclear receptors to promote gene expression and protein synthesis is one of several variations on the regulation of gene expression. The process by which nitric oxide relaxes smooth muscle has several features in common with the visual process in the retina of the eye, as well as with other physiological processes. The means by which insulin and other growth factors affect cellular metabolism will be described in the context of obesity (chapter 14). Research is just beginning to reveal how these few common signaling mechanisms can give rise to a wide range of biological responses. This has been a brief and incomplete description of the actions of messengers in the body, but it is intended to give the reader an appreciation for how signals affect their target cells and for the role of lipids as messengers.

5

Oxidation and Lipid Peroxidation

Oxygen is essential for survival. Without sufficient oxygen, as a result of suffocation or because the heart stops pumping oxygenated blood through our arteries, we can die within minutes. When life in the form of simple unicellular organisms evolved, there was virtually no oxygen in the earth's atmosphere. It was not until photosynthetic organisms such as cyanobacteria evolved that oxygen began to accumulate in the atmosphere and in the oceans. However, because of its chemical reactivity, it was rapidly consumed in the early environment. Substances with which it reacted were ultimately used up (oxidized), and oxygen began to accumulate. It eventually made up about 20 percent of the atmosphere, where it has remained for hundreds of millions of years (Kump, Kasting, and Crane 1999, 182).

Oxygen's Toxic Properties

Primitive biological organisms floating around in the early seas, when oxygen levels were low, needed to develop ways to protect themselves from oxygen and its highly damaging free radicals. Those that did not evolve with protections succumbed to the ravages of oxygen in their environment and died out. Most organisms evolved with elaborate systems, programmed in their genetic makeup, to ward off the detrimental reactions of oxygen. Eventually some organisms evolved with systems to utilize oxygen for their own benefit; these are known as aerobic organisms. They tend to be much more complex than anaerobic organisms, including their ways for handling oxygen to avoid being destroyed by this life-sustaining gas.

Substances that readily react with oxygen are generally known as reduced substances, whereas those formed after they react with oxygen are oxidized substances. The process of reacting with oxygen is known as oxidation. Chemists have several ways of defining oxidation, but it will suffice to say that substances that are produced after reactions with oxygen are oxidized, whether they react directly with oxygen or indirectly through intermediary molecules.

Oxygen may go through several stages in the process of reacting with other substances, depending on what substance it reacts with and the environment where the reaction takes place. The stages that oxygen goes through between molecular oxygen in the air and the combined form in water, as a result of a series of single-electron reductions ($+e^-$), are represented as follows:

$$O_2 \xrightarrow{+e^-/H^+} O_2^-/HO_2 \xrightarrow{+e^-/H^+} H_2O_2 \xrightarrow{+e^-} OH^{\bullet} \xrightarrow{+e^-/H^+} H_2O \ / \ H_3O^+ \ / \ OH^-$$

where O_2 is molecular oxygen; O_2^- is known as superoxide anion (a free radical); HO_2 is the perhydroxyl radical (an acidic form of superoxide); H_2O_2 is hydrogen peroxide, a highly reactive substance but not a free radical; OH (often written as $OH^{\bullet}$) is the hydroxyl radical, one of the most reactive substances known, which never exists for more than a fleeting instant in the environment because of its propensity to react with almost anything it encounters; and H_2O is water, which also exists in its acidic (H_3O^+) and alkaline (OH^-) forms. It is possible for hydrogen peroxide to form its alkaline anion (HO_2^-), although this is important only under strong alkaline conditions.

Another form of molecular oxygen, known as singlet oxygen ($^1O_2$), is much more reactive toward organic and biological molecules than the common form of oxygen in the atmosphere, which is technically known as triplet oxygen ($^3O_2$). Singlet oxygen may be generated in biological systems under certain conditions and may be involved in many of the damaging reactions discussed below. The free radicals and other reactive forms of oxygen listed here are collectively known as reactive oxygen species. Ozone and some highly reactive organic substances (such as lipid peroxides discussed below) are also lumped into this category of reactive oxygen species. Although they may occasionally play a beneficial biochemical or physiological role, they are more often associated with adverse health conditions. Their detrimental chemical characteristics are frequently involved in toxic effects.

Managing Our Oxygen

When we breathe air into our lungs, oxygen crosses membranes of cells lining the alveoli of the lungs, enters the blood, and crosses membranes of red blood cells in the capillaries of the lungs. Red blood cells contain hemoglobin, a protein whose primary function is to escort oxygen to tissues where it is needed and prevent unwanted reactions. Hemoglobin is uniquely designed to pick up oxygen molecules in the lungs, where oxygen is abundant, and release it in tissues that are actively working. Muscle has another specialized protein known as myoglobin, which binds oxygen after it is delivered and stores it for times when muscle cells suddenly need the oxygen. Myoglobin binds oxygen even more strongly than hemoglobin does, so when the tissue has used up most of its oxygen reserves, myoglobin draws the oxygen away from hemoglobin—much like a game of tug-of-war—to replenish the stockpile in that tissue. Consequently, these two proteins keep most of the oxygen supply bound up in their clutches, with very little oxygen floating around freely to react indiscriminately with substances in the body.

In order for cells to utilize oxygen for their metabolic needs, the oxygen must be able to diffuse into cells to specific sites where proteins ensure that the proper chemical reactions take place. This means that the oxygen must break free from hemoglobin and myoglobin to traverse cell membranes. There are occasional mishaps, and free oxygen can react with substances in cells that lead to destructive consequences. Destructive effects caused by oxygen in biological tissues are known as oxidative stress. When this happens, cellular enzymes can often repair the damage done and keep cells functioning as they should. If there are too many incidents of oxygen reacting with cellular components, the damage may become too extensive for repair processes, and cells may die. Many cells in a tissue may succumb to the same oxidative stress, resulting in tissue necrosis (death of the tissue).

It is fortuitous that oxygen does not react with most biological or organic molecules under normal physiological conditions (body temperature and neutral pH). However, there are catalysts that can speed up random reactions of oxygen with biological molecules. For example, ferrous iron (Fe(II) or Fe^{2+}) is very good at promoting reactions of oxygen with biological molecules. Organisms evolved mechanisms early in evolution to avoid having free iron floating around in cells; virtually all the iron in living organisms is bound up in proteins, either to impart a specific function to that protein, as in hemoglobin and myoglobin, or to transport or store the iron. Transferrin is a protein that transports iron through the blood, and ferritin is another protein in cells that stores iron after it has been delivered to cells. When proteins that need iron are synthesized, just enough iron is provided, and the rest remains bound to ferritin. There are also proteins that bind copper and other metals to prevent those metals from catalyzing unwanted reactions involving oxygen.

Although oxygen and metal ions are bound up by proteins and their free concentrations are severely limited, it is necessary for them to diffuse from one binding site to another for their proper utilization. During these fleeting instances reactions may occur that result in damage. It is necessary to get free oxygen, a catalyst, and the organic molecules together in the same place for reactions to occur. Free oxygen is especially prevalent near the lungs. If substances such as low-density lipoprotein (LDL) are in the circulation for a long time, there is greater likelihood that conditions will be met for oxidation reactions to ensue. There are environmental pollutants (ozone) and foreign substances (xenobiotics) that can stimulate these deleterious reactions.

Antioxidant Systems

Organisms evolved with numerous systems to protect them from the ravages of oxygen. In addition to proteins that keep oxygen and metal ion catalysts bound up and unreactive, there are enzymes that eliminate reactive oxygen species, such as superoxide radical and hydrogen peroxide. Most organisms make the enzyme superoxide dismutase (SOD) to remove superoxide because it can do damage, but, more importantly, it may be involved in producing stronger, more destructive oxidizing agents. Superoxide dismutase removes two superoxide radicals in each catalytic cycle, transforming them into one molecule of oxygen and one molecule of hydrogen peroxide.

All cells have enzymes for destroying hydrogen peroxide, either by reducing it to water (these are called peroxidases) or by a dismutation of two hydrogen peroxide molecules to form one molecule of oxygen and two molecules of water (figure 5.1). Catalase is the enzyme that transforms hydrogen peroxide by dismutation and is among the fastest enzymes known; it catalyzes millions of reactions per second when high levels of hydrogen peroxide are present.

Hydroxyl radical is so reactive that there does not appear to be an enzyme for eliminating it. This renegade reacts with almost everything it encounters and therefore would not accumulate in the environment of a cell to make it necessary to have an enzyme for its elimination. Instead, cells have antioxidant molecules that may be in relatively large abundance to intercept this highly reactive radical and minimize damage to other cellular components if it does get formed.

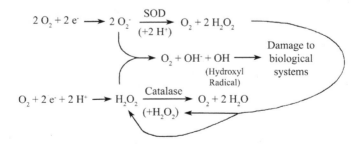

FIGURE 5.1. Superoxide anion reacts with hydrogen peroxide to produce hydroxyl radical. The enzymes superoxide dismutase (SOD) and catalase remove superoxide (O_2^-) and hydrogen peroxide (H_2O_2), respectively, forming innocuous products of oxygen and water.

Ozone, which may enter the body from polluted air, is also highly reactive, and there does not appear to be an enzyme for its elimination. Vitamin E and vitamin C are antioxidants that inactivate these highly reactive oxygen species, but there are others. Those derived from plant sources are known as phytochemicals, such as carotenoids and flavonoids (some are lipids and some are water-soluble). There are also systems that evolved to repair damage when protective systems fail. First, let us look at some of the consequences to cell components if the protective systems fail, especially those involving lipids.

Normal Metabolic Oxidation of Lipids

It was mentioned (chapter 2) that triglycerides are storage fats that can be used for energy. In order for the energy to be utilized from these triglycerides, the fatty acids are first released from the glycerol that holds them. The fatty acids are then metabolized by enzymes that break them down to carbon dioxide and water, consuming oxygen in the process. This metabolic process is known as the fatty acid beta-oxidation pathway. The oxidation reactions take place within the mitochondria of cells and are carefully orchestrated to make sure the oxygen is utilized in just the right way and the energy is used to make ATP (adenosine triphosphate). ATP is the energy currency in a cell, much as electricity is an energy currency in our commercial infrastructure and money is an economic currency in our financial affairs. These currencies are generated for the purpose of sustaining the needs of their respective systems.

The major energy supplies in the diet come from macronutrients—carbohydrates, fats, and proteins. These organic nutrients undergo controlled oxidation through several series of indirect reactions with oxygen to produce carbon dioxide as the ultimate oxidized form of carbon. Mitochondria are special organelles in aerobic cells that are very efficient at producing ATP through controlled oxidation reactions. Cells use ATP for all their necessary energy-requiring functions, including synthesis of new cell components, bringing most nutrients into the cells, communicating with other cells, or receiving signals from hormones and other messengers, to name just a few. The mitochondria are much like a power plant in a city that produces electricity, where the electricity is used to construct new buildings in the city from raw materials, for transport of people and supplies, and for communication within the city and with other cities.

Practically all fatty acids can be oxidized by the mitochondrial beta-oxidation pathway. Several studies have shown that different categories of dietary fats are metabolized in the body at different rates. Oleic acid (a major monounsaturated fatty acid in foods, especially olive oil) gets metabolized for energy at a faster rate than either stearic acid (a long-chain saturated fatty acid) or linoleic acid (a major polyunsaturated fatty acid in vegetable oils) (Jones, Pencharz, and Clandinin 1985). Saturated fat is not absorbed as well from the intestinal tract, which only partially explains the approximately fivefold difference in rate of oxidation of oleic acid compared with stearic acid. Only about 15 percent of the absorbed oleic acid was oxidized in the first nine hours after eating, compared with about 10 percent of linoleic acid and 3 percent of stearic acid. These low percentages of fatty acid oxidation would suggest that subjects were depositing much of the dietary fat in tissues, since they were confined to a metabolic ward while the measurements were made and consequently not utilizing much energy for exercise. Fats, as well as sugar, are used for energy in muscle tissue during exercise.

Lipid Peroxidation

There is another possibility for oxidation of polyunsaturated fatty acids, such as linoleic acid. This oxidation does not rely on enzymes to orchestrate a controlled metabolic pathway but instead results from arbitrary or random reactions involving oxygen and generates many products that can be detrimental to cells. It also does not produce ATP. Catalysts are usually involved to cause oxygen to react with molecules with which it would not normally react. The reactions involve reactive oxygen species, such as superoxide, singlet oxygen, hydrogen peroxide, ozone, and hydroxyl radical. They are often referred to as free radicals, although some, like hydrogen peroxide, are not technically radicals. Like political radicals, they want to change things in their environment. Some molecules are more easily attacked than others by free-radical molecules, just as bullies in a school yard find it easier to attack some kids more than others. Cells usually try to avoid forming free radicals, or utilize them only for specialized reactions that require this kind of activation.

With regard to fatty acids, the polyunsaturated fatty acids (PUFAs) are easily attacked by free radicals, whereas saturated fatty acids are completely resistant and monounsaturated fatty acids are moderately resistant to attack. When a free radical encounters a PUFA, it extracts a hydrogen from the PUFA, preferably from the $-CH_2-$ between the double bonds (as seen in figure 5.2). The free radical ends up as a stable, nonradical entity (HRad in figure 5.2), and the PUFA becomes a free radical (indicated by the dot in figure 5.2). In other words, the free-radical chain of events is propagated. Free oxygen will in turn readily react with the PUFA radical. The resulting product is another radical, but with a molecule of oxygen (O_2) now attached. This radical then abstracts a hydrogen atom from another PUFA, generating a new PUFA radical, and becoming what is known as a fatty acid peroxide—similar to hydrogen peroxide but one hydrogen is replaced by the fatty acid. Just as hydrogen peroxide is very reactive and is used as an antiseptic (to kill bacteria), the fatty acid peroxides are also very reactive and can do damage inside cells.

This spontaneous or nonmetabolic oxidation of fatty acids generally takes place in the phospholipids of cell membranes and lipoproteins and is known as lipid peroxidation because lipid peroxides are formed in the process. Once

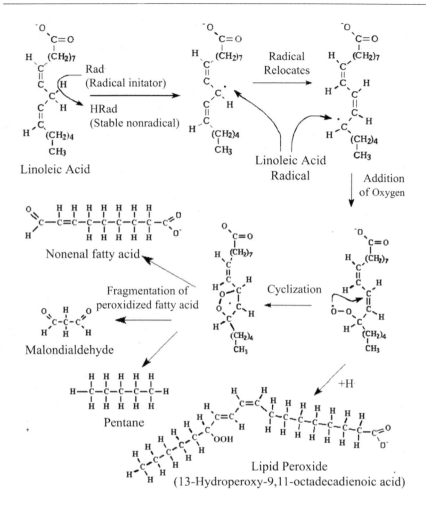

FIGURE 5.2.   Scheme for lipid peroxidation of linoleic acid. Rad• is a free-radical initiator, and several end products include a nonenal fatty acid, malondialdehyde, pentane, and 13-hydroxy-9,11-octadecadienoic acid, a lipid hydroperoxide.

lipid peroxidation gets started, it can continue until terminated by protective mechanisms in the cell. The membranes are more prone to this type of oxidation because they are exposed to oxygen entering cells from the blood and being distributed to organelles (cellular compartments surrounded by membranes) within the cells for metabolic reactions.

One protective mechanism used by cells is interception of radicals by vitamin E incorporated in membranes. Vitamin E is a lipid or fat-soluble vitamin and an antioxidant because it prevents the oxidation (or peroxidation) of other lipids, such as PUFAs in phospholipid membranes. Vitamin E actually sacrifices itself to the free radicals, becoming a free radical in the process. The vitamin

E free radical is not nearly as reactive as the fatty acid free radicals, so it is not on a mission to react with other molecules. Eventually, it will react either with another vitamin E radical or with enzymes that restore vitamin E to its active status in membranes. In other words, vitamin E is regenerated or recycled.

Vitamin C is another antioxidant in cells, but it is a water-soluble vitamin and intercepts radicals in the watery matrix (cytoplasm) of cells rather than in the lipid membranes. Vitamin C and vitamin E work together to intercept free radicals and protect both the lipid and aqueous parts of cells from harmful oxidation processes.

There are also enzymes that catalyze the breakdown of fatty acid peroxides. These enzymes are known as peroxidases. One of them, glutathione peroxidase, contains the mineral selenium, which accounts for most of our nutritional need for this trace element. The peroxidase enzymes generally convert fatty acid peroxides to molecules that are much less reactive.

When peroxidation has taken place in a cell membrane, there are enzymes to repair the damaged site. An oxidized PUFA will be removed and replaced by a fresh fatty acid, perhaps another PUFA, restoring the membrane to its viable state. If the peroxidative damage is too extensive and repair enzymes are unable to keep up with the oxidative reactions, the membrane can leak electrolytes, resulting in cell death. This is like having a hole in a boat. If the hole is small, it may be possible to remove water fast enough to stay afloat. But if the hole is too large, water may enter the boat faster than the bailer can remove it, and the boat will sink.

Cell survival is dependent on keeping electrolytes such as sodium outside the cell and potassium inside. This mutual exclusion of electrolytes is maintained by what is known as the sodium-potassium pump, a membrane protein that moves sodium from inside the cell to the outside while moving potassium from outside to inside. This requires energy, just like pumping water, and the energy is provided by ATP generated in metabolic reactions. When membranes spring a leak, sodium and potassium can float freely through the membranes, and the sodium-potassium pump will try to restore the vital electrolyte balance. It may use up all the energy stores (ATP) of the cell trying to do so.

Reactive Oxygen Species in Cancer and Heart Disease

If oxidative stress is extensive but the cell manages to survive, the free radicals may disrupt several cell systems (see figure 5.3). If they manage to get to the nucleus of the cell and cause oxidation of DNA, this would amount to genetic mutations. Again, there are special proteins to protect the DNA and enzymes to repair DNA if it has been oxidized. However, these systems can also be overwhelmed if free radicals are too numerous. If mutations in DNA are not repaired and these occur in critical sites, they may create an aberrant cell. Such mutated cells may be recognized by surveillance cells of the immune system and be destroyed. If they evade the immune surveillance, they may become tumorous progenitors (cancerous cells). As cancerous cells divide, producing more and more aberrant cells that do not respond to normal control signals, they eventually become a cancerous tumor. The chances of all these events happening are extremely small, but they only have to happen once in the trillions of cells in the body during a lifetime in order for cancer to develop.

This is just one scenario for carcinogenesis. There are many other possibilities for cancer development, but not all of them involve peroxidation of PUFAs. However, lipid peroxidation is believed to be a major factor in the development of cancer. In the early years of research on the links between dietary fats, blood cholesterol levels, and risk of heart disease, it was found that substituting PUFAs (from vegetable oils) for saturated fats in the diet decreased cholesterol in the blood but increased chances of developing cancer (Pearce and Dayton 1971). There will be more about the role of polyunsaturated fatty acids in cancer in chapter 10.

As mentioned in chapter 2, lipids are transported through the blood by lipoproteins, which are more or less spherical entities with triglycerides, cholesterol, and other lipids packaged in the interior of the sphere, while proteins and phospholipids make up the outer shell of the lipoprotein. It is possible for PUFAs in the phospholipids of lipoproteins, especially LDL (low-density lipoprotein),

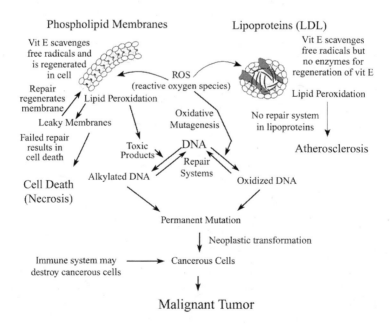

FIGURE 5.3. The consequences of oxidative stress on cell membranes, lipoproteins, and DNA.

to encounter free radicals and undergo spontaneous oxidation (as seen on the right side of figure 5.3). LDL may also become oxidized by vascular endothelial cells (lining the arteries) or by other cells in the circulation (Steinberg 1997). Because the LDL is not a cell, it does not have resident enzymes to repair or remove lipid peroxides. Our bodies have evolved a system to remove the entire oxidized LDL particle in lieu of repairing it. Deposition of oxidized LDL in the lining of the arteries, especially around the heart, leads to atherosclerosis, which is described in more detail in chapter 6.

For decades, doctors have recognized that people with high levels of LDL, not just cholesterol, will be at higher risk for heart disease. High blood cholesterol was considered to be the original risk factor, but once lipoprotein particles could be differentiated, LDL was found to be the culprit. LDL is the lipoprotein with the highest cholesterol content (nearly 50 percent cholesterol and cholesteryl esters; see appendix D). Clinical laboratories usually do not measure LDL directly but use a formula involving the level of triglycerides, total cholesterol, and HDL-cholesterol to calculate an estimated level of LDL-cholesterol. Generally, an individual's risk for heart disease is based on the ratio of total serum cholesterol to HDL-cholesterol.

HDL contains only about 25 percent cholesterol and cholesteryl esters, and may help to remove cholesterol from atherosclerotic deposits. HDL also seems to be less susceptible than LDL to oxidation in the circulation. In fact, HDL was found to have its own antioxidant and anti-inflammatory properties (Barter et al. 2004). The greater susceptibility of LDL to lipid peroxidation is probably due to the greater amount of time that individual LDL particles remain in circulation, which increases the probability they will encounter the conditions necessary for lipid peroxidation to be initiated. There will be more on this in the following chapters.

Conclusions

Although oxygen is vital for our normal metabolism of macronutrients (carbohydrate, fat, and protein) to produce carbon dioxide and energy, it is also capable of producing undesirable chemical reactions in the body. These anomalous reactions can destroy vital cell components and compromise the viability or integrity of cells. If these undesirable oxidative reactions become too extensive, it can result in a condition known as oxidative stress, which may lead to tissue necrosis (death of cells) or other forms of permanent damage. If DNA becomes oxidized or otherwise modified by some of the reactive products of lipid peroxidation, this can lead to cancer. We will see in the following chapters that lipid peroxidation, the oxidative destruction of polyunsaturated fatty acids in lipids, especially cell membranes and lipoproteins, is responsible for several life-threatening diseases.

PART TWO

Dietary Fats in Health and Disease

6

Atherosclerosis

There seems to be a pervasive popular belief that cholesterol clogs the arteries, with cholesterol buildup eventually becoming so great that blood flow to the heart or brain becomes blocked, causing heart attack or stroke. Biomedical research over the past fifteen to twenty years has shattered this model of cardiovascular disease. Indeed, half of all myocardial infarctions (blood clots in the heart) occur in arteries with less than 50 percent narrowing from atherosclerotic buildup, which is considered to be only mild atherosclerosis (Smith 1996). Scientists now realize the subtle complexities of atherogenesis (development of atherosclerosis) and the life-threatening complications that arise as a consequence.

The simplistic model invoked gradual deposition of cholesterol along the lining of the arteries, much like the deposition of scale and corrosion in pipes. Metal pipes carrying water through a hot-water heater may accumulate minerals to the extent that water flow becomes so restricted that little water can pass through them. This was a convenient, mechanistic representation for atherosclerotic plaque buildup and the resulting heart attack. However, our vascular system is a little more complicated than even the most sophisticated plumbing systems. Cholesterol is not causing atherosclerosis and heart disease but is being dragged into the fray by a complex cast of molecular characters.

Arteriosclerosis was recognized in the nineteenth century as a vascular disease that could lead to death from heart attack or stroke. It was known that arteries became ossified, a process similar to bone formation, hence the term "hardening of the arteries." In the 1850s the noted physiologist Rudolf Virchow described atherosclerosis as an inflammatory disease of the arterial inner layer (Libby 2000). This condition seemed to progress with age and was generally looked upon as one of the inevitable maladies of aging. Fatty deposits along the arterial lining were known to have a high content of cholesterol, in addition to fats and minerals such as calcium and phosphate. It is interesting that the concept of ossification, or bone formation in the arteries, fell out of vogue in the twentieth century. It was replaced by the notion that calcium and phosphate simply precipitated in the fatty deposits—much like scale buildup in hot-water pipes.

The simplistic view of passive deposition of these artery-clogging fats and minerals was probably more appealing than trying to invoke active biological processes, since the molecular and cellular events that resulted in atherosclerotic plaque were not understood. Historical developments that led to the hypothesis that cholesterol was responsible for atherosclerosis and heart disease were discussed in chapter 1. The Framingham Heart Study found a greater incidence of heart disease among people with high blood cholesterol levels, although high blood pressure, arrhythmias, and cigarette smoking also increased risk of heart disease. Several independent studies showed that saturated fats in the diet would raise blood cholesterol levels, while polyunsaturated vegetable oils would lower blood cholesterol. Today the biochemical mechanisms for the cholesterol-lowering effect of polyunsaturated fatty acids are known (see chapter 4).

Whether a person has high or low serum cholesterol is genetically determined. Many people with high cholesterol or, more specifically, high levels of LDL (low-density lipoprotein) have a genetic predisposition for this condition (Sing and Boerwinkle 1987; Knijff et al. 1994; Eichner et al. 2002). Some individuals exhibit large fluctuations in serum cholesterol with changes in their diet, while others experience relatively little change in cholesterol levels in response to the same dietary changes—again there is probably a genetic link. Diet cannot be ignored as an important factor contributing to one's overall health, well-being, and longevity. The modern "American" diet may be having a profound negative influence on vascular as well as other diseases.

This chapter will discuss the biochemical and physiological bases for atherogenesis (development of atherosclerotic plaque) and myocardial infarction, and how dietary factors, specifically polyunsaturated fatty acids, influence these processes. There are several factors that are good indicators of heart disease risk (discussed in chapter 7); serum cholesterol, or more specifically, LDL-cholesterol, is only one of them. However, LDL is the major factor in promoting atherosclerosis, yet cholesterol in the LDL particles has very little to do with it.

A Changing Paradigm of Atherogenesis

Several decades ago it was recognized that fat deposits are formed by a specialized group of white blood cells known as macrophages that become lodged in the space between endothelial cells, which form the lining of arteries, and smooth-muscle cells surrounding the arteries. The latter contract and relax in their autonomic rhythms, giving rise to the peristaltic action for pumping blood through the arteries. The tube that forms the blood vessels is made up of vascular endothelial cells held together by fibrous proteins such as collagen.

Circulating mononuclear leukocytes (one type of white blood cell also known as monocytes) manage to squeeze through the arterial walls at specific sites in response to intercellular signals. They take up residence in the arterial intima—the space between the endothelial cells and smooth-muscle cells.

There are many signals sent and received by cells involved in this process. It suffices to say that vascular cells, especially around the heart for individuals prone to heart disease, secrete proteins known as adhesion molecules, such as vascular cell adhesion molecule-1. These cell adhesion proteins bind strongly to proteins on the surface of circulating blood monocytes that have been activated by oxidized or modified LDL. When the monocytes stick to these vascular cells on the walls of the artery, they are able to slip through the vessel walls. Other signaling molecules known as cytokines or chemoattractants are also involved in provoking this response. Several of these cytokines have been identified and have been found to be made by specific cells in the area as part of an elaborate array of events that are not completely understood yet (see figure 6.1).

Specific substances in oxidized LDL particles can trigger events that get cells to produce cytokines and other signaling molecules that stimulate removal of oxidized LDL from the circulation and promote its deposition in atherosclerotic lesions surrounding the arteries. As indicated in the previous chapter, it is the

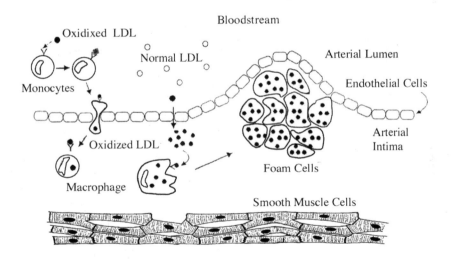

FIGURE 6.1. Atherogenesis. Oxidized LDL particles bind to receptors on monocytes, triggering them to bind to vascular endothelial cells and move into the arterial intima. The monocytes are transformed into macrophages that engulf oxidized LDL particles once they are in the arterial intima, until they are filled with lipids and appear as foam cells. The buildup of foam cells is known as an atheroma or fatty streak. Such fatty streaks can eventually become calcified to produce atherosclerotic plaque.

Adapted from Thijssen and Mensink 2005.

polyunsaturated fatty acids (PUFAs) in phospholipids of the LDL particle that are most susceptible to oxidation.

Daniel Steinberg and colleagues at University of California, San Diego, have shown that people who consumed diets rich in linoleic acid (polyunsaturated) had LDL that was far more susceptible to oxidation than people fed oleic acid–rich (monounsaturated) diets (Reaven et al. 1993). Peroxides of linoleic acid, the major PUFA in vegetable oils, have been shown to stimulate the expression of some of the signaling proteins involved in promoting the removal of oxidized LDL from the circulation (Inoue et al. 2001). Judith Berliner and colleagues at University of California, Los Angeles, have identified several other oxidation products of phospholipids in LDL particles (see figure 6.2) and found that they trigger production of adhesion molecules in vascular cells (Navab et al. 2004). This causes the monocytes to stick to those cells and take up residence in the arterial intima (Subbanagounder et al. 2000). Fructose oxidation products and homocysteine (another important risk factor for heart disease) may also modify LDL particles to stimulate their removal from the circulation by arterial macrophages.

Most cells remove cholesterol from the blood by synthesizing an LDL receptor that resides on the cell membrane and binds to normal (nonoxidized) LDL particles. The cue for synthesis of the receptor is a low level of cholesterol in the cell. When cholesterol levels are high inside the cell, the amount of receptor on the plasma membrane will decrease, and less LDL will be taken up. On the other hand, removal of modified or oxidized LDL from circulation is carried out by a limited number of specialized cells that make scavenger receptors for modified or oxidized LDL. The cells that make these receptors include arterial macrophages and perhaps smooth-muscle cells lining the arteries. There have been a few types of scavenger receptors identified, each showing a preference for a particular type of modification. LDL may also aggregate as a result of certain types of enzyme activities that lead to rapid uptake by the scavenger system (Steinberg 2002).

Once monocytes take up residence in the arterial intima, they become known as macrophages—a white blood cell with a specific mission. These macrophages are very efficient at engulfing oxidized or modified LDL particles. They engorge so much modified LDL that, under the microscope, they look as if they are filled with foam; hence the descriptive name of foam cells. It was naturally assumed that these cells were scavengers for circulating cholesterol, but it was later recognized that they have a preference for the major lipoprotein that carries cholesterol, namely low-density lipoprotein. LDL happens to be composed of about 45 percent cholesterol and cholesteryl esters. The vascular macrophages have no appetite for normal LDL but show a keen discrimination for oxidized or chemically modified LDL, which resembles oxidized phospholipids on bacterial cells—another quarry of macrophages.

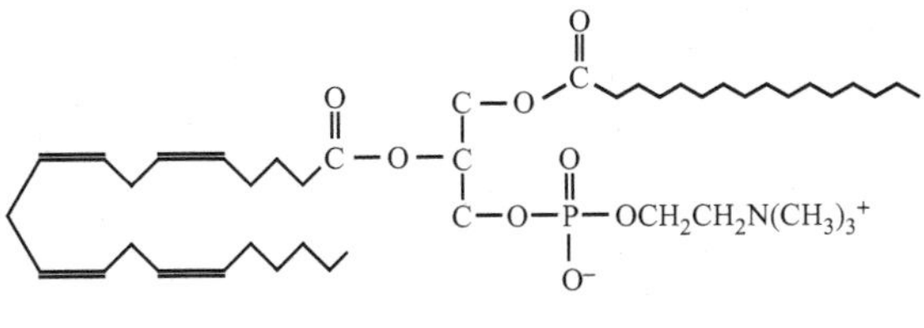

1-Palmitoyl-2-arachidonyl-sn-glycero-3-phosphorylcholine
a phospholipid found in normal LDL particles

Oxidation of archidonate fatty acid group
(may be initiated by lipid peroxidation)

Identified Inflammatory Products of Oxidized LDL

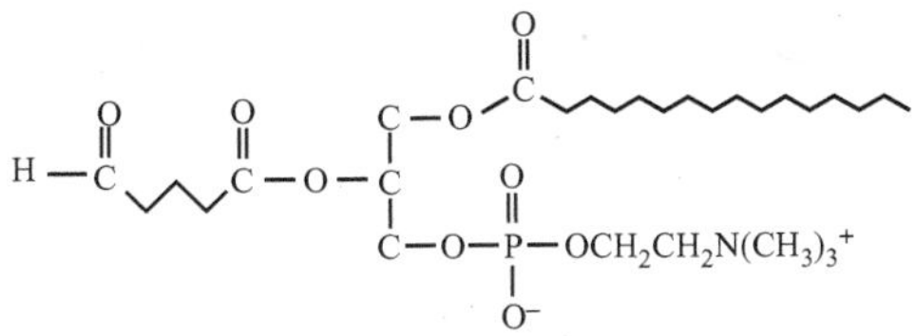

1-Palmitoyl-2-oxovaleryl-sn-glycero-3-phosphorylcholine
(POVPC)

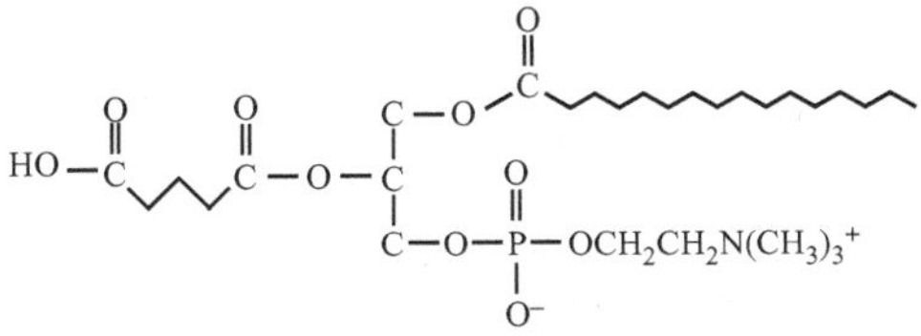

1-Palmitoyl-2-glutaryl-sn-glycero-3-phosphorylcholine
(PGPC)

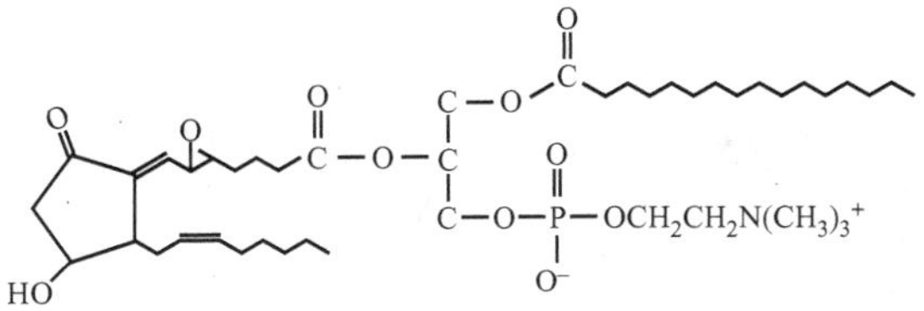

1-Palmitoyl-2-(5,6-epoxyisoprostane E$_2$)-sn-glycero-3-phosphorylcholine
(PEIPC)

FIGURE 6.2. Products formed by oxidation of arachidonate in phospholipids of LDL. The three oxidation products featured here have been shown to stimulate the expression of adhesion proteins, such as vascular cell adhesion molecule-1, resulting in monocytes binding to endothelial cells and working their way into the arterial intima (see figure 6.1). These oxidized phospholipids have also been shown to be proinflammatory mediators (see text).

Adapted from Navab et al. 2004.

There has been a major focus on LDL as the blood component that causes heart disease, although most people, including a majority of health professionals, have overlooked the fact that it is usually oxidized polyunsaturated fatty acids in LDL phospholipids that are actually causing these lipoproteins to be deposited in the lining of arteries. The composition of LDL particles varies relatively little with changes in cholesterol consumption (Kummerow et al. 1977), although changes in the ratio of dietary saturated fatty acids to PUFAs has a greater influence on cholesterol levels and LDL (see chapter 1). More importantly, the quantity of vegetable oil in the diet will profoundly affect the amount of PUFAs in those particles, which would make them more susceptible to oxidation.

Oleic acid found in canola oil and olive oil is much less prone to oxidation than linoleic acid, which is abundant in soy-, corn-, sunflower-, and safflower-based vegetable oils. Saturated fatty acids are completely resistant to oxidation and other chemical modifications. If there is relatively little linoleic acid present in LDL, the chances of it getting oxidized would be small, whereas if LDL is loaded with PUFAs from eating mostly vegetable oils, it is much more likely that some will get oxidized. Furthermore, once oxidation gets started, it is like the flu, being passed from one PUFA to another if they are numerous and in close contact with one another (see chapter 5).

Counteracting the susceptibility of PUFAs to oxidation is the effect of PUFAs on the activity of normal LDL receptors and how fast most cells in the body remove LDL from circulation before it gets oxidized. When someone consumes greater amounts of polyunsaturated oils and lesser amounts of saturated fats, the normal LDL receptors are more effective at removing LDL from the blood, whereas a low level of polyunsaturated oils in the diet results in less active receptors (Woollett, Spady, and Dietschy 1992). This means that LDL will be removed from the blood faster when a person consumes more polyunsaturated oils, which accounts for the lower LDL-cholesterol levels associated with such a diet. In addition, there is a slower production of LDL in the liver when there are more polyunsaturated fatty acids in the diet. Consequently, when there is less LDL in the circulation and it remains there for a shorter time, there would be less chance for it to undergo oxidation or other modifications that earmark it for scavenger removal by the arterial macrophages.

The beneficial effect of dietary polyunsaturated vegetable oils lowering serum cholesterol and the detrimental effect of them making LDL more susceptible to oxidation are most likely waging a constant battle with one another in the body. It is difficult to say which of these effects will overshadow the other. Any number of unknown influences may determine whether oxidation susceptibility or more rapid removal of LDL will be the dominating factor in any individual, which would likely be determined by genetics as well as other dietary components, whether the person smokes, and exercise.

Vegetable oils have antioxidants in them, such as vitamin E. High temperatures used in frying foods can cause vegetable oils to begin oxidizing, and vitamin E may be depleted in oils used in frying. Consequently, fried vegetable oils would contain lower levels of antioxidants and would also have some oxidized PUFAs. It is not known just how much of the oxidized fatty acids in fried food get absorbed and end up in LDL, since there are many stages in the absorption and mobilization of dietary fats where these oxidation products are eliminated. It seems likely that, even with protective mechanisms, some of the oxidized fatty acids might get absorbed and packaged into lipoproteins like LDL.

There have been conflicting reports regarding whether vitamin E supplements are beneficial in suppressing oxidation of LDL and subsequent deposition in atherosclerotic plaques (Cynshi and Stocker 2005; Pryor 2000). There is a general consensus that vitamin E should be beneficial, but it may depend on whether it is naturally in food or whether it is given in a purified form, as in a dietary supplement. When taken as a supplement, it may not get incorporated into unsaturated oils very well, especially if supplements are taken without food or taken with foods that have relatively little fat.

One would expect omega-3 PUFAs in fish oils to be as susceptible to lipid peroxidation as omega-6 PUFAs in vegetable oils, if not more so, since omega-3 PUFAs tend to have more double bonds. It was mentioned in chapter 5 that multiple double bonds make fatty acids more susceptible to oxidation. However, omega-3 PUFAs have been found to be less susceptible to lipid peroxidation than their omega-6 counterparts (Visioli, Colombo, and Galli 1998).

Omega-3 fatty acids have been shown to protect against heart disease by lowering blood pressure and heart rate, reducing inflammation, and improving vascular and platelet functions (Mori and Woodman 2006). Although these favorable effects are generally attributed to eicosapentaenoic acid (EPA), docosahexaenoic acid (DHA) has been found to improve cardiac arrhythmias and reduce clotting. These effects are far more significant than the cholesterol-lowering properties of omega-3 PUFAs. Some omega-3 fatty acids seem to suppress lipid peroxidation and behave as antioxidants (Richard et al. 2008).

Atherogenesis as a Protective Mechanism

The cascade of cell responses triggered by the presence of oxidized LDL in the bloodstream appears to be an elaborate system for removing oxidized LDL from the circulation as a protective measure. Although continuous buildup of these scavenged lipoprotein particles in the arterial intima may cause life-threatening complications in the long run, it may spare the body from the more immediate threat of these toxic lipids and their degradation products. When our ancestors evolved, the amount of PUFAs in the diet would have been much less than the amount in a typical Western diet today. This should be acutely obvious when perusing the vegetable oil aisle of any supermarket in North America.

In the early 1980s Guy Chisolm and coworkers at the Cleveland Clinic Foundation in Ohio showed that reactive oxygen species (such as superoxide radical and hydrogen peroxide) are involved in the formation of oxidized LDL particles (Morel, Hessler, and Chisolm 1983). It was known at that time that LDL could be cytotoxic, in other words, can damage cells and tissue under certain conditions. They found that antioxidants and free-radical scavengers could prevent formation of the cytotoxic form of LDL during preparation of the lipoprotein particles for cell culture experimentation. However, adding antioxidants and radical scavengers to (human fibroblast) cell cultures and then administering previously oxidized LDL did not prevent cell injury. They concluded that oxidized lipids in the LDL particles were the toxins that were doing damage to the cells.

Since lipid peroxide levels in plasma (Sato et al. 1979) and serum lipoproteins (Nishigaki et al. 1981) had been shown to be elevated in diabetic patients, the results of Chisolm and coworkers explained why diabetics were much more prone to atherosclerosis and heart disease. More recent studies have shown that fructose, one of the sugars produced by the enzymatic hydrolysis of sucrose (commercial sugar) but also a major component of high-fructose corn syrup, which is widely used in American beverages and foods, can generate free radicals and damage proteins at a much faster rate than glucose (blood sugar) (Sakai, Oimomi, and Kasuga 2002). Fructose also produced lipid peroxides in human LDL at a rate several times faster than glucose in an incubation medium. Such results suggest that the dietary choices prominent in the United States, namely vegetable oils with abundant polyunsaturated fatty acids and high-fructose corn syrup as a sweetener, are providing the ideal conditions to promote atherosclerosis.

There is still a question of whether oxidation of PUFAs in phospholipids of LDL particles is primarily caused by certain cells in the body or whether it is simply a spontaneous chemical reaction catalyzed by substances floating freely in the bloodstream. LDL particles do get oxidized in the body (Palinski et al. 1989). It was found that white blood cells placed in a petri dish can oxidize PUFAs in LDL when they are activated by certain substances that elicit an immune response (Quinn et al. 1987). Oxidized LDL can damage vascular endothelial cells (Rong et al. 1998), which may promote further oxidation of LDL as a result of the immune system response to the vascular damage (white blood cells can generate free radicals in the process of cleaning up debris from damaged vascular tissue). It is still not clear whether oxidative processes in the body or oxidation prior to ingestion accounts for the bulk of oxidized LDL in the circulation, although both may very well be contributing to the problem.

PUFAs in the phospholipids surrounding LDL particles make up a relatively small percentage of the particles, yet are probably responsible for most of the havoc wrought. By the mid-1990s it had become clear that cholesterol was being dragged into the arterial macrophages because it happened to be in LDL particles containing oxidized PUFAs. Chisolm and coworkers isolated several components of oxidized LDL and tested them on human fibroblast cell cultures to determine which ones were cytotoxic (Chisolm et al. 1994). One of the highly toxic components they found was cholesterol hydroperoxide. Considering the nature of their assay for cytotoxicity, it is quite possible that the cholesterol hydroperoxide was initiating peroxidation of membranes in the cells used in the assay. Cholesterol hydroperoxide was also found to be present in human atherosclerotic lesions, meaning that it was not just an artifact of the cell culture medium.

Although this would indicate that cholesterol may also be forming toxic products, the amount of the cholesterol hydroperoxide formed appears to be small compared with PUFA oxidation products, and cholesterol peroxidation is not easily initiated. Richard Lerner's group at Scripps Research Institute in La Jolla showed that ozone formed in atherosclerotic lesions can react with cholesterol to produce several cholesterol oxidation products (Wentworth et al. 2003). So cholesterol oxidation may be the result of free radicals and reactive oxygen species generated in atherosclerotic lesions. These oxidations involve white blood cells mounting an attack on abnormal lipids and lipoproteins in an attempt to eliminate them. There has not been as clear a demonstration of cholesterol oxidation products stimulating the transformation of monocytes as there has been for peroxides of linoleic acid and other forms of oxidized phospholipids (Navab et al. 2004). However, cholesterol and its esters are in the core of LDL particles, whereas phospholipids are on the surface.

The cytotoxic effects of lipids in circulating LDL particles are perhaps setting in motion an immune response to clean up these damaged cells and remove the modified lipoproteins. The leukocytes that are responsible for this cleanup may be generating reactive oxygen species and free radicals in the process of eliminating damaged cells and their debris, which could further stimulate oxidation of lipids in circulating LDL. These events may be triggering damage to the fibrous cap that ultimately leads to a heart attack (see below).

Progression from Fatty Streaks to Atherosclerotic Plaques

Accumulation of a significant number of foam cells gives rise to the appearance of fatty streaks. This preliminary buildup may be reversible, and there is good evidence that apolipoprotein E (ApoE) or apolipoprotein A-I (ApoA-I), components of high-density lipoprotein (HDL), can help remove cholesterol and other lipids from foam cells. This is believed to be a major reason why HDL is considered "good" cholesterol. Although details of the process are not known, the ApoE and ApoA-I are most likely entering the arterial intima (as shown in figure 6.3), where they bind to receptors on foam cells. These apolipoproteins are the protein components of the lipoproteins, which are lacking lipids that get packed in the interior of the lipoprotein particle. You can think of the apolipoproteins as being like shopping carts that you fill with groceries when you go to the supermarket. The lipid-free apolipoproteins seem to go around the body looking for lipids that have been sequestered, particularly along the lining of the arteries. Proteins, such as cholesteryl ester transfer protein, may help move cholesteryl esters and cholesterol to these apolipoproteins, ultimately assembling a complete HDL particle. HDL particles then find their way back into the circulation, carrying cholesterol and other lipids to cells throughout the body to be used for productive purposes.

Oxidized lipids may not get transferred to ApoE and ApoA-I, and may remain and accumulate in the atherosclerotic lesions. At some point, accumulation of oxidized lipids in the fatty streaks triggers formation of a fibrous cap over this debris (see figure 6.4). Linda Demer and coworkers at University of California, Los Angeles, found that inflammatory lipids in oxidized LDL (shown in figure 6.2), and inflammatory cytokines produced in response to oxidized lipids, can promote calcification (mineralization) of atherosclerotic lesions (Abedin, Tintut, and Demer 2004). Oxidized LDL stimulates leukocytes to produce platelet activating factor (PAF), which is an unusual phospholipid with many potent signaling effects. Platelet activating factor stimulates expression of cytokines, such as tumor necrosis factor-alpha (TNF-α), which signals vascular cells to become transformed into calcifying vascular cells. The scenario is complicated, with a large number of cytokines and other signaling factors involved. The Demer group also found that DHA, a major component of omega-3 fish oils, can suppress vascular calcification through its actions on the peroxisome proliferator-activated receptor alpha (PPARα) (Abedin et al. 2006).

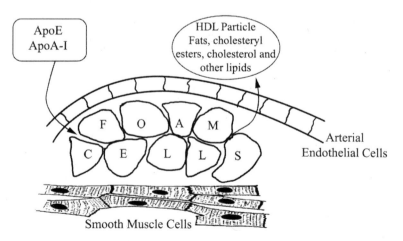

FIGURE 6.3.   Foam cells in fatty streaks are accessible to apolipoproteins (e.g., ApoA-I and ApoE), which can remove cholesteryl esters, cholesterol, and other lipids. Transfer proteins move lipids from the foam cells to the apolipoproteins, forming complete HDL particles in the process, which reenter the circulation and deliver these lipids to cells throughout the body.

Calcifying cells share characteristics of bone-forming cells known as osteoblasts and produce extracellular matrix molecules, which act as a scaffold of proteins on which to build the fibrous cap. In addition to connective proteins, such as collagen and elastin, there is osteopontin, the same fibrous phosphoprotein found in bones, which binds calcium in an array that promotes mineralization. Expression of the enzyme alkaline phosphatase is also stimulated. This enzyme releases phosphate from organic sources for bone formation or, in the vascular case, for mineralization of atherosclerotic plaque.

There is a relationship between aortic calcification and osteoporosis, which is independent of age. It seems the body will remove calcium from bone in order to form calcified deposits around atherosclerotic plaques. Demer speculates that this mechanism may have evolved in parallel with the body's response to chronic infection. Microbes destroyed by the immune system are rich in oxidized lipids resembling oxidized LDL, and if the immune system is not able to completely eliminate this debris, the connective tissue responds by forming scar tissue or a calcified barrier around it (Demer and Tintut 2003).

Calcification of arteries may not always lead to life-threatening events, such as heart attack or stroke, but seems to be a normal process in formation of atherosclerotic plaque. Not all fatty streaks along the linings of arteries appear to be calcified, and just what determines whether a fatty streak will become calcified is still not well understood. High-fat diets promote calcification, whereas lipid-lowering drugs (statins) inhibit calcification. Patients receiving statin therapy after a myocardial infarction had much greater coronary artery calcification than patients on statin drugs who did not have a previous infarction, even though LDL levels were essentially the same in both groups (Callister et al. 1998). The underlying reason for some patients having greater calcification

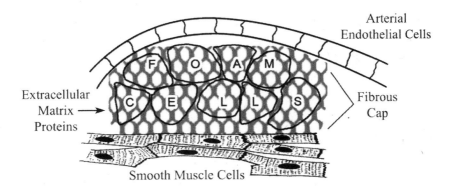

FIGURE 6.4. Lipid oxidation products induce transformation of vascular cells to calcifying-type cells, resulting in production of extracellular matrix proteins that bind calcium and phosphate to produce bonelike mineralized scar tissue over the oxidized lipid deposits. The fibrous cap makes the fatty deposits inaccessible to apolipoproteins.

than others was not clear. Perhaps calcification of the coronary arteries is what made one group prone to myocardial infarction in the first place. Coronary calcification also shows a positive correlation with levels of LDL and an inverse correlation with levels of HDL. Consequently an undesirable serum lipid profile is promoting calcification of the fatty deposits that ultimately leads to dangerous conditions for heart disease complications.

Although the events in cap formation are just beginning to be understood, there is good evidence that rupture of the fibrous cap can lead to myocardial infarction and other life-threatening events. The cells involved in bone resorption, osteoclasts, are actually derived from leukocytes and are a special type of macrophage—cousins of the foam cells. One can think of them as eating or consuming mineral deposits, and they may be responsible for some of the damage to calcified deposits that lead to a heart attack or stroke (as shown in figure 6.5).

Chronic infections, especially by *Chlamydia pneumonia*, have been implicated as a cause of heart attacks. It seems likely that the inflammation triggered by the immune system mounting an attack on this bacterial infection may trigger a cascade of events that leads to a blood clot or myocardial infarction. Lipopolysaccharides on the surface of many bacteria, when they come in contact with white blood cells, cause the leukocytes to produce a host of messengers, ranging from bioactive eicosanoids derived from PUFAs to cytokine proteins that whip the immune system into action to try to destroy these invaders. Activated leukocytes may inadvertently attack components of the fibrous cap overlying atherosclerotic lesions, resulting in exposure of the fibrous proteins (collagen). Such exposure activates circulating blood platelets, triggering clot formation. If there is an abundance of arachidonic acid available (from high levels of vegetable oil in the diet) for eicosanoid production, there could be

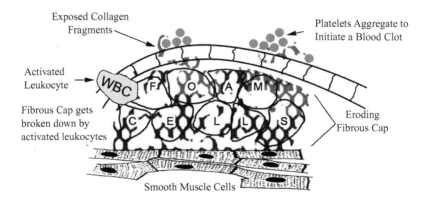

FIGURE 6.5. Infiltration by activated leukocytes releases proteases and breaks down the fibrous cap, exposing collagen to circulating platelets, which will trigger platelet aggregation and initiation of a blood clot.

greater activation of leukocytes and platelets, resulting in a more profound or widespread inflammatory response and consequent clotting. Unfortunately, some individuals have had the stage set to lead this performance into a tragic ending.

Life-Threatening Complications of Atherosclerosis

The buildup of atherosclerotic plaque does not necessarily cause heart attacks or myocardial infarction, although increased amounts of atherosclerosis does make myocardial infarction more likely. Indeed, the amount of arterial obstruction by atherosclerotic buildup has little relation to the likelihood of blood clot formation, but disruption of the fibrous cap covering artherosclerotic plaques triggers a blood clot. It is not so much the quantity of plaque buildup but rather the quality of the plaque. A key event in clot formation results from platelets coming in contact with collagen, the protein that holds cells together. The majority of myocardial infarctions seem to be caused by a rupture of the fibrous cap, which exposes collagen and other substances that trigger platelets to initiate a blood clot.

It appears that inflammatory mediators produced by leukocytes in the vicinity of atherosclerotic plaques, whether because of bacterial infections or other causes, can result in rupture of the fibrous cap. One of these mediators is interferon-gamma, which may interfere with the ability of smooth muscle to repair and maintain the cap. The rupture process may involve release of enzymes into the space surrounding cells, which causes some breakdown of proteins in the fibrous cap. Again, there appears to be a host of proinflammatory mediators released that trigger rupture of the fibrous cap and lead to clot formation. Ultimately, circulating platelets will produce thromboxane A_2 (TXA_2), an eicosanoid formed from arachidonic acid, which sets everything in motion for clot formation (see figure 6.5). As mentioned earlier, if there is more arachidonic acid available because large amounts of omega-6 polyunsaturated vegetable oils are constantly being consumed in the diet, there could be more extensive blood clotting and a more massive heart attack.

Enzymes involved in breaking down the fibrous cap, known as proteases (or more specifically metalloproteases), are normally regulated by protein inhibitors that are produced by certain cells in response to specific signals. Such inhibitory proteins have been identified in human atherosclerotic plaques. It is not yet clear whether changes in levels of these inhibitors are involved in rupture of the cap. There are also large numbers of immune cells, such as T lymphocytes and macrophages (nonfoam cells) involved. The macrophages may be of the osteoclastic type, which would break down calcified plaque deposits. There also seem to be fewer smooth-muscle cells in regions where there are ruptured fibrous caps that have led to thrombosis. It is believed that some events may have triggered these muscle cells to die via programmed cell death, or apoptosis.

Numerous studies indicate the degree of vascular blockage by atherosclerotic buildup is not as important as other factors in setting the stage for myocardial infarction or other life-threatening cardiovascular events (Libby 2000). Disruption of the fibrous cap covering the plaque, whether by mechanical fracture (due to a bodily jolt or increased pressure) or by chemical erosion, can expose components that trigger formation of a blood clot. Chemical erosion can be due to release of metalloproteases that gradually break down the molecular architecture (see above). Many of the molecular events that lead to these physical complications are triggered by lipids—not benign lipids, such as cholesterol and saturated fatty acids, but powerful bioactive agents formed from PUFAs. It stands to reason that having a greater abundance of these PUFAs in cell membranes, particularly the omega-6 variety, would result in an exaggerated response. This could mean the difference between a fatal heart attack and a mild one.

The working model for atherosclerosis and life-threatening cardiovascular events is changing dramatically, but the world has not been quick to adapt. Peter Libby of Harvard Medical School wrote that a new understanding of the processes involved in atherogenesis could "reduce reliance on costly technologies that address the symptoms rather than the cause of atherosclerosis" (Libby 2000). Fats are playing complicated roles in all this. The saturated fats can raise levels of LDL, but oxidation of polyunsaturated fatty acids in the LDL particles makes them destined for scavenging by the arterial macrophages to produce atherosclerosis. Which is the greater culprit in this physiological drama? The monounsaturated fatty acids, such as those in olive, canola, and palm oils, seem to have escaped indictment on all accounts. On the other hand, omega-6 poly-unsaturated oils, such as corn, soy, safflower, and sunflower oils, are wreaking havoc in our bodies, including the vascular system. The food industry seems to be working on developing grains and oil sources with more favorable fatty acid profiles. We can hope they will bring them to market in the near future.

The new model might appear to have one major problem: if cholesterol is not the cause of heart disease, why is it a marker for the risk of getting this disease? The medical establishment honed its skills in measuring serum cholesterol and can give patients a simple printout of their blood work that shows the levels of total cholesterol, HDL-cholesterol, and the ratio of these two along with calculated values for LDL-cholesterol. These lipoprotein profiles are typically being used by physicians to predict a person's risk for heart disease. A great deal of time, money, and effort has been spent trying to come up with correlations between serum cholesterol levels, lipoprotein ratios, and heart disease risk, yet more than half the people having myocardial infarctions and strokes have low or normal cholesterol levels (Libby 2000). Is a more aggressive pharmaceutical approach to lowering blood cholesterol in the general population the answer to healthier and more productive lives? Some of these issues will be discussed in the following chapters.

Conclusions

As research continues on these complex processes, it is becoming clear that the cholesterol hypothesis is outdated. Serum total cholesterol and LDL-cholesterol are only surrogate indicators of disease risk. The focus placed on lowering cholesterol through dietary intervention with vegetable oils does not alter the death rate from all causes (see chapter 1). The beneficial effects of the cholesterol-lowering drugs are most likely due to their wide range of pleiotropic effects, which will be discussed in chapter 8. Many of the manifestations of heart disease are accompanied by inflammatory events in the vascular system, and the role of polyunsaturated fatty acids in inflammation will be discussed in chapter 9.

The United States still remains one of the leading nations for incidence of heart disease, even though immense effort and money have been invested to counter this troubling statistic. Diabetes, obesity, and metabolic syndrome are compounding heart disease, and diet is clearly involved in promoting all of these. Not only dietary fats but sugars as well are playing major roles in the increased risk for heart disease. The dietary advice of avoiding saturated fats and cholesterol seems to be focusing too much on readily measurable markers and ignoring the profound effects that polyunsaturated vegetable oils can have on health and longevity, many of which will be discussed in later chapters.

7

Risk Factors
in Cardiovascular Disease

There has been much made of the association between dietary saturated fats and high serum cholesterol levels, particularly LDL-cholesterol. Although high serum cholesterol was identified as a major risk factor for heart disease in the Framingham Heart Study, several international studies have not found a good correlation between serum cholesterol and death from all causes. Generally, there is a correlation between serum cholesterol and coronary death rate. It is important to keep in mind that the prevalence of death from coronary heart disease rises rapidly with age, particularly above seventy years of age. Often there is some correlation between death from all causes and serum cholesterol in younger populations (less than seventy years), as in the American and Finnish cohorts of the Seven Countries Study (Keys 1980, 124), but that correlation disappears as the cohort ages (Menotti et al. 2001). The American and Finnish cohorts had the greatest incidence of coronary deaths in the Seven Countries Study, which influenced the correlation between serum cholesterol and the all-cause death rate in those groups.

This chapter will look at a variety of other risk factors for heart disease, many of which appear to be much better predictors of heart attack than serum cholesterol. In many cases, these additional risk factors, such as inflammation, interact with high levels of LDL or serum cholesterol to produce oxidized LDL, which is responsible for atherogenesis. Some of these risk factors may promote erosion of atherosclerotic plaque, which leads to myocardial infarction, as described in the previous chapter.

Inflammation as a Predictor of Heart Attack

During the 1990s researchers found that a substance known as C-reactive protein (CRP) in the blood was much better than serum cholesterol levels as a predictor of heart attacks. Mark Pepys and colleagues at the Royal Free and University

College Medical School in London found that heart attack patients who had high levels of CRP in the blood were much more likely to have a bad outcome than those with lower levels of CRP (Koenig and Pepys 2002). The Pepys group also showed there was a linear relationship between levels of CRP in blood and risk of heart disease. Paul Ridker and colleagues at Harvard Medical School in Boston found that blood CRP level was a useful predictor for first heart attack or stroke in patients, especially postmenopausal women, and that CRP and cholesterol levels were not related to one another; one could have high cholesterol and low CRP levels or vice versa (Willerson and Ridker 2004).

It was also found that one of the benefits of daily aspirin was directly related to CRP and that statin drugs may, but do not necessarily, lower CRP levels. CRP levels are also linked to obesity (as measured by body mass index). If a person loses weight, the levels of CRP go down. CRP is also higher in people with type 2 diabetes, and high CRP levels have been associated with syndrome X (or metabolic syndrome), which increases risk of diabetes, myocardial infarction, and stroke (Willerson and Ridker 2004). Metabolic syndrome is characterized by the presence of excess abdominal fat, insulin resistance, increased blood pressure, and increased levels of glucose and triglycerides in the blood, as well as lower levels of HDL—in other words, an unhealthy state.

Moderate physical activity is generally considered to be beneficial in reducing one's risk of heart disease. It was found that participants in the ATTICA study in Greece who engaged in physical activity had significantly improved levels of many inflammatory risk factors relative to their sedentary counterparts. Exercisers exhibited a 29 percent decrease in CRP and 20 percent decrease in tumor necrosis factor-alpha (TNF-α), while white blood cell counts of exercisers were only about one-fifth (19 percent) those of the sedentary group (Panagiotakos et al. 2005). Levels of TNF-α and CRP were also found to be elevated in prehypertensive patients relative to individuals with normal blood pressure. Hypertension is another important risk factor in heart disease.

Depression was also associated with both inflammation and thrombosis factors in people without active cardiovascular symptoms. Elevated inflammation indicators may explain the increased frequency of cardiovascular events in depressed individuals (Panagiotakos et al. 2004). A report from the Augsburg, Germany, cohort of the World Health Organization (WHO) MONICA project showed that interaction of high CRP levels and depressed mood in individuals resulted in a highly significant increase in risk of myocardial infarction (Ladwig et al. 2005). Depression alone will often increase the incidence of heart attacks, although the trend is not always enough to show a significant impact (Alboni et al. 2008). In the same Augsburg cohort of the MONICA project, it was found that obesity alone was not a strong predictor of heart attack but depression in obese individuals became a very strong predictor (Ladwig et al. 2006).

Although CRP is a good predictor of heart disease, little is known about its role; is it a cause or merely another marker produced in response to other underlying complications? It is generally accepted that elevated levels of CRP are caused by the inflammatory response. However, CRP can augment the inflammatory response by causing an increase in production of reactive oxygen species and inflammatory cytokines, and an increase in expression of cell adhesion molecules (Jialal, Devaraj, and Venugopal 2004). CRP can also increase the uptake of LDL by macrophages, which could be due to a higher rate of LDL oxidation taking place because of an inflammatory condition (associated with activated leukocytes). CRP may enhance blood clotting and was shown to increase the size of blood clots in rats when they were induced to have a myocardial infarction.

A recent large-scale study treated apparently healthy men and women who had low serum LDL-cholesterol (less than 130 mg/dL) but elevated CRP (greater than 2 mg/L) with rosuvastatin (a cholesterol-lowering statin drug) or placebo (Ridker et al. 2008). The study was terminated after a median of less than two years of follow-up because of the dramatic benefits of the drug compared with a placebo. Statin therapy lowered LDL-cholesterol by 50 percent but also lowered CRP by 37 percent. There were significant reductions in myocardial infarction, stroke, revascularization (coronary bypass surgery), and unstable angina in the statin-treated patients relative to the placebo control group. Unlike the profile of total serum cholesterol, LDL-cholesterol, and HDL-cholesterol that is routinely given in blood analyses, levels of CRP are usually not measured unless a physician or patient requests it. If the above study had not been focused on CRP, the striking improvement in outcome for the statin-treated group would be interpreted as being due to the large, drug-induced decrease in LDL-cholesterol of patients with low or normal levels of LDL-cholesterol.

CRP is thus another interesting player in the story of cardiovascular diseases, although it seems this blood protein is produced by the liver in response to cytokine signals from immune cells. It has been found that genetic variations (polymorphisms) in human CRP influence the basal levels of CRP in the circulation as well as CRP production in response to other stimulatory factors (Brull et al. 2003). Higher levels of CRP may be an inherent property of inflammatory conditions in the body and, as such, may end up being a secondary player as far as causing heart disease is concerned. Nevertheless it seems to be a much more useful indicator of heart disease risk than serum cholesterol. There is also an association between CRP and metabolic syndrome and diabetes (Ridker et al. 2003).

Diabetes Increases Risk of Vascular Diseases

Diabetes is a major risk factor for heart disease and stroke, as well as several other maladies. Elevated levels of sugar in the blood can result in glycosylation of proteins (glucose or similar carbohydrate derivatives become covalently bound to proteins). Glycation of proteins in circulating LDL, and possibly VLDL, results in removal of the modified LDL particles by macrophages lining the arteries. This occurs via the scavenger receptors for LDL, as described in the previous chapter. Sugars, such as glucose and fructose, bind to a wide range of proteins in the body, and when they are at elevated levels in the circulation, they are more likely to react. Fructose is about ten times more reactive than glucose in terms of protein glycation (Sakai, Oimomi, and Kasuga 2002). Furthermore, if there is inflammation, the sugars may be oxidized more easily by free radicals generated in the inflammatory response, resulting in greater binding to proteins.

According to the American Diabetes Association, two out of three people with diabetes die from heart disease and stroke. In general, there are increased levels of markers of oxidative stress, lipid peroxidation, and inflammation (such as CRP) in patients with type 2 diabetes (el-Mesallamy, Suwailem, and Hamdy 2007). It should not be surprising that diabetes causes a dramatic increase in risk of cardiovascular diseases. The role of dietary sugars in obesity, diabetes, and metabolic syndrome is described in more detail in chapter 14.

The Genetic Connection

It has long been recognized that a predisposition to cardiovascular diseases is hereditary. More recently it was realized that high levels of cholesterol in the blood may also be hereditary and that a genetic predisposition to high serum cholesterol outweighs any dietary fat contribution with respect to cardiovascular disease. Michael Brown and Joseph Goldstein received the Nobel Prize in Physiology or Medicine in 1985 for their pioneering work on the genetic basis for familial hypercholesterolemia, a disease characterized by extremely high levels of LDL in the blood and consequent heart disease at a young age (Brown and Goldstein 1986). Their research showed that people inheriting one copy of a defective gene (heterozygous) typically have levels of plasma LDL-cholesterol around 300 mg/dL (about three times the normal level), and if a person has the rare misfortune of inheriting two copies of a defective gene (homozygous), plasma levels of LDL-cholesterol can be as high as 650 mg/dL (or about six times normal levels). They identified several different genetic mutations that all lead to disruption of the structure and function of the LDL receptor on cells. Therefore, cells with such receptors are not able to remove LDL from the circulation for a long time, where much of it gets oxidized. The oxidized LDL is then removed by vascular macrophages. In their study, subjects in the homozygous population had severe atherosclerosis in their teenage years.

Brown and Goldstein have recently described several other genetically determined diseases that result in elevated levels of plasma cholesterol because of defective removal of LDL from the bloodstream (Goldstein and Brown 2001). In these cases, the defective proteins exhibit little or none of the activity of their normal counterparts. In addition to defective LDL receptors, there can also be defective apolipoprotein B-100 (apoB-100), which is the protein in

LDL particles that binds to the LDL receptor. Gross changes in the structure of either the LDL receptor or apoB-100 as a result of genetic mutations can result in these proteins not recognizing one another. Consequently, the LDL particles circulating in the blood do not bind to their receptors on cells and cannot be taken into most cells that would normally utilize them. This would be like trying to rake leaves with only the handle of a rake because of a manufacturing defect that resulted in the head of the rake falling off or never getting attached.

For a biochemist, it is easy to imagine a wide range of genetic defects, not only in the LDL receptor and apoB-100 on LDL particles, but in many other proteins that are needed to control levels of cholesterol in the blood. These may include proteins responsible for regulating the synthesis of cholesterol in cells and other proteins involved in transporting LDL particles across cell membranes and processing them inside the cells. For example, one such genetic defect has been identified that results in a defective protein that normally functions to bring the LDL particle into a cell after it binds the LDL receptor on the cell surface (Mishra, Watkins, and Traub 2002). The genetic mutations need not produce proteins that are completely nonfunctional but may produce proteins that just do not work as efficiently as they should. As an example, the substitution of one amino acid for another in the LDL receptor may result in apoB-100 binding, but not as tightly as it normally would, resulting in twice as much time being needed to remove LDL from the blood. This would be analogous to raking the lawn with many teeth of the rake missing. The lawn may eventually get cleaned, but it will take a longer time.

There has also been much interest in the size of LDL particles circulating in the blood. Smaller LDL particles are associated with a higher risk of atherosclerosis than larger LDL particles. Smaller LDL particle size is not only associated with more cardiovascular disease but also correlates with elevated levels of triglycerides in the blood and decreased levels of HDL-cholesterol (Campos et al. 1992). Individuals with larger LDL particles have lower overall levels of LDL-cholesterol as a result of fewer LDL particles. One study found that small LDL particle size increased the relative risk of myocardial infarction by a factor of nearly seven times (Griffin et al. 1994). On the other hand, individuals in families whose members have a high probability of living past the age of one hundred have significantly larger HDL and LDL particle sizes compared with members of control families with normal life spans (Barzilai et al. 2003). They also have lower prevalence of hypertension, cardiovascular disease, and metabolic syndrome, and increased levels of a particular gene for cholesteryl ester transfer protein (see below).

HDL has been separated into various subgroups or subpopulations based on size, density, and electrophoretic behavior, with HDL-2 and HDL-3 representing two major subpopulations that are distinguished by ultracentrifugation. Electrophoresis of HDL has been used to obtain several subpopulations designated as alpha-1, alpha-2, alpha-3, prealpha-1 to prealpha-3, prebeta-1, and prebeta-2.

Researchers in Boston found that heart disease was associated with alpha-3 and prealpha-1 HDL particles in male participants of the Framingham offspring study group (see chapter 1). These HDL particles have small diameters. They also found a negative or inverse correlation with larger alpha-1 and prealpha-3 HDL particles (Asztalos et al. 2004). These results confirm many other studies that have found an inverse correlation between HDL particle size and risk of heart disease.

ApoA-I is the major apolipoprotein found in HDL particles. High-fat diets have been shown to increase levels of ApoA-I by increasing its production and decreasing its rate of degradation or hydrolysis by proteases. Levels of ApoA-I are controlled mainly by its rate of degradation, more so than by its rate of production (Asztalos et al. 2000). HDL particle size is also dependent on its rate of degradation. Genes that code for several different proteins, such as lipoprotein lipase, hepatic lipase, lecithin:cholesterol acyl transferase, cholesteryl ester transfer protein, and phospholipid transfer protein, influence the size of HDL particles.

A rare genetic disorder known as Tangier disease is characterized by accumulation of cholesteryl esters in several tissues, including the tonsils, adenoids, liver, and spleen. Patients with the homozygous form of this disease have extremely low levels of HDL, whereas heterozygotes (inheriting one normal and one defective gene) have below-normal HDL-cholesterol and smaller HDL particle size (Brousseau et al. 2000). The genetic defect in Tangier disease is known to affect a protein called ATP-binding cassette-1 (ABC-1), which plays a role in removal of cholesterol from atherosclerotic lesions, although its precise role is not well understood. Genetic variations in this gene other than those responsible for Tangier disease would likely also affect the levels and size of HDL particles.

Genetic variations (polymorphisms) in ApoE also affect transfer of lipids to HDL particles containing this apolipoprotein and, therefore, efflux of cholesterol from atherosclerotic deposits back to the circulation. There is much interest in these ApoE polymorphisms with respect to heart disease risk, but it suffices to say there are many possible genetic variations that have an effect on the size and abundance of both HDL and LDL and, consequently, on the risk of heart disease. Another possible genetic trait influences the level of homocysteine in the blood.

Homocysteine Affects Several Systems
Involved in Cardiovascular Disease

Homocysteine is an amino acid related to methionine and cysteine. It is an intermediate in the conversion of these sulfur-containing amino acids to one another and in their metabolism in the cell to other products. Damage to vascular cells by homocysteine has been attributed to several possible mechanisms, including oxidative stress (production of hydrogen peroxide, superoxide, and other reactive oxygen species). Homocysteine inactivates nitric oxide, which is a beneficial messenger that can dilate blood vessels to lower blood pressure and exerts a wide range of other physiological effects, as described in chapter 4. Homocysteine inhibits glutathione peroxidase, which removes toxic lipid peroxides by reducing them to inactive alcohols. It also inhibits glutathione synthesis, which influences a wide range of beneficial effects attributed to this biological antioxidant (McCully 1998).

Elevated levels of homocysteine in blood increase the likelihood of thrombotic events through multiple effects on several clotting factors, most likely through addition of homocysteine to many of the proteins involved—homocysteine has a propensity to react with protein (D'Angelo and Selhub 1997). Homocysteine can modify LDL, making this lipoprotein susceptible to removal from the circulation by vascular macrophages and thereby promoting atherosclerosis. Because homocysteine has a free sulfhydryl group, it is also easily oxidized by oxygen and may result in generation of free radicals and reactive oxygen species that can wreak havoc with cell membranes, LDL particles, and other blood components.

A recent meta-analysis of twenty-six reports found that there was a significant positive association between homocysteine and heart disease (Humphrey et al. 2008). The meta-analysis revealed that for each 5 micromole per liter increase in plasma homocysteine there was a 20 percent increase in the chance of having a cardiovascular event. Normal homocysteine levels should be less than about 10 micromole per liter. Although homocysteine is not a strong predictor of heart disease and stroke, it may interact with other risk factors, such as serum LDL or C-reactive protein.

Kilmer McCully of the Department of Veterans Affairs Medical Center in West Roxbury, Massachusetts, suggested that the declining rate of heart disease in the United States since the 1960s may be attributable to increased intake of vitamin supplements, particularly vitamin B_6 and folate, which form important coenzymes in homocysteine metabolism (McCully 1998). This decline may also have been influenced by greater availability of fresh fruits and vegetables in Western markets during winter months. The genetic disease known as homocysteinuria is most commonly caused by an inherited deficiency of the vitamin B_6-dependent enzyme cystathionine synthase (Mudd, Skovby, and Levy 1985). In many cases, vitamin B_6 supplementation could alleviate symptoms of the disease.

Homocysteine levels appear to be another useful indicator for risk of heart attacks and stroke. Dietary factors, in this case vitamins B_6, B_{12}, and folate, may have some influence on homocysteine levels. Unfortunately homocysteine, like CRP, is not routinely analyzed when blood samples are drawn. If a patient is diagnosed with elevated homocysteine, dietary supplements of these vitamins would certainly be warranted.

Conclusions

Clearly there are many genetic factors involved in raising or lowering one's risk for heart disease, and there are many measurable factors in the blood that can be used as predictors. High levels of serum cholesterol (greater than 240 mg/dL) and LDL-cholesterol (greater than 150 mg/dL) are good measures of risk for cardiovascular disease, but C-reactive protein (CRP) may be a better indicator. CRP is associated with inflammation, which facilitates atherosclerosis as well as promoting the disruption of atherosclerotic plaque to initiate thrombosis and myocardial infarction. The biochemical role of CRP is just beginning to become clear. Diabetes and its associated complications may work in conjunction with inflammation and is another important risk factor.

Heredity is much more important as a risk factor for heart disease than the profile of fatty acids in the diet. This is not to say that people should not strive to eat a well-balanced diet with ample amounts of proper nutrients, but attempts to eliminate or minimize foods that contain saturated fats and cholesterol, per se, may not be the best approach. The polyunsaturated fatty acids found in vegetable oils may exhibit beneficial effects with respect to blood cholesterol levels, but they are also clearly involved in a wide range of effects that can exacerbate many diseases, including arthritis and other types of inflammation, atherosclerosis, heart disease, and stroke. The following chapters will describe more diseases that are affected by these essential fatty acids.

8

Lipid-Lowering Drugs

Akira Endo was not yet a teenager when Alexander Fleming won the Nobel Prize in Physiology or Medicine in 1945 for his discovery of antibiotics from the mold *Penicillium notatum* in 1928. Endo was fascinated by Fleming's work and got a job with the Sankyo Chemical Company isolating enzymes from fungi and molds for processing fruit juices while he was pursuing his Ph.D. in the early 1960s. In 1966 he went to Albert Einstein College of Medicine in New York to work for two years on cholesterol, which was becoming a major topic of research interest at that time (Landers 2006).

After Endo returned to Sankyo, he convinced his supervisor to allow him to search for fungal substances that would inhibit cholesterol synthesis. He and his colleague Masao Kuroda tested thousands of fungal broths in the early 1970s before finding one that worked on the key enzyme in cholesterol synthesis, 3-hydroxy-3-methylglutaryl coenzyme A reductase (HMG-CoA reductase). They found that *Penicillium citrinum* produced inhibitor(s) of the enzyme and eventually isolated 23 mg of a crystalline material from about 600 liters of culture (Endo 1992). By the end of 1973 they had determined the structure of the active substance, and it became known as mevastatin, since it suppressed the formation of mevalonate, the product of the reaction catalyzed by HMG-CoA reductase. The same substance was isolated from *Penicillium brevicompactum* in England in 1976 and named compactin (Brown et al. 1976). The drug is variously known as ML-236B, mevastatin, and compactin.

When Endo's group tested their mevastatin on rats, they found it could decrease plasma cholesterol at high doses (20 mg/kg), but the effects were not reproducible. They tried feeding rats a diet supplemented with 0.1 percent mevastatin for a week, but it resulted in no changes in plasma cholesterol. A chance meeting with a colleague provided an opportunity to study the effects of their new inhibitor in a different animal—female chickens. Feeding the hens a diet containing 0.1 percent mevastatin resulted in about a 50 percent reduction in plasma cholesterol, without affecting diet consumption, body weight, or egg production. Experiments with dogs and monkeys showed it worked in those species as well. They eventually found that rats and mice had an unusual ability to produce several-fold higher levels of HMG-CoA reductase when receiving multiple exposures to the inhibitor, although inhibition was effective with a single acute dose (Endo 1992).

In 1977 Endo approached Akira Yamamoto, a physician who was treating patients with familial hypercholesterolemia at Osaka University Hospital in Japan. Endo provided some mevastatin to try on one patient with extremely high levels of plasma cholesterol, beginning with a dose of 500 mg per day. After two weeks the plasma cholesterol level decreased by about 20 percent, but levels of creatine phosphate and transaminase were elevated, and the patient developed muscular weakness (Endo 1992). The adverse effects disappeared when the drug was withdrawn. The drug was then given at 200 mg per day without adverse effects on muscle. There was little detectable change in plasma cholesterol, but after five months the xanthomas (lipid-filled nodules under the skin) and other noticeable traits of the disease were markedly reduced. Preliminary tests in other patients with familial hypercholesterolemia and hyperlipidemia also showed reductions in plasma cholesterol levels over several months at various doses.

These early studies were done before internal review boards and other strict regulations became commonplace in clinical trials. When managers at Sankyo saw the results Yamamoto had with hypercholesterolemic patients, they decided to test mevastatin in formal clinical trials. Endo was offered an academic position at Tokyo Noko University. Sankyo management was not pleased with Endo's departure, and the company rarely mentioned him in its publicity about the discovery of the first statin drug. Endo received recognition for his discovery by winning several awards and prizes, including the 2008 Albert Lasker Medical Research Award.

In 1976 the Sankyo Chemical Company entered into an agreement with the pharmaceutical giant Merck and Company, providing Merck with samples and unpublished data regarding mevastatin. Merck scientists repeated the work with laboratory animals and achieved similar results. They soon isolated a mevastatin analog from another fungus, *Aspergillus terreus*, which they named lovastatin (Endo 1992). Lovastatin was slightly more active than mevastatin at inhibiting HMG-CoA reductase. Endo also had independently isolated lovastatin in his first year at Tokyo Noko University, but Merck held patent rights to the compound in the United States. Merck obtained approval from the U.S. Food and Drug Administration to market lovastatin as Mevacor in 1987. It was the first FDA-approved statin to be marketed.

Mevastatin was originally isolated as the lactone (closed-ring structure) form of the acid and alcohol groups in the molecule (see figure 8.1). It was later found that the open-ring acid form, which is more water-soluble, is a more potent inhibitor of HMG-CoA reductase. The drug competes with HMG-CoA, the natural substrate, for binding to the active site of the enzyme. It is a reversible and competitive inhibitor of the enzyme. It was found that the acid form of mevastatin binds to the enzyme with an affinity about ten thousand times greater than that of HMG-CoA (Endo 1992).

During the 1980s several similar drugs were discovered, either by isolation from fungi or through chemical synthesis and modification of natural products. The statins all have the 3-hydroxyglutaryl group in common, which has a chemical similarity to HMG-CoA, as shown in figure 8.1. This allows them to compete with the natural substrate for binding to the active site, while the bulky chemical groups attached to 3-hydroxyglutaryl can vary quite a lot. The bulky groups attached probably impart their differences in in terms of potency or affinity for binding the enzyme.

Statins lower cholesterol levels in the body by inhibiting a key enzyme, 3-hydroxy-3-methylglutaryl coenzyme A reductase (HMG-CoA reductase), in the pathway for cholesterol synthesis. When this pathway for cholesterol synthesis is shut down, or at least slowed down substantially, the body needs to get its cholesterol either from the diet or from storage sites in the body—the latter being limited but often significant. In addition to cholesterol stored in adipose tissue (fat stores), there is cholesterol in the fatty streaks lining the arteries, as described earlier regarding the development of atherosclerosis. When the liver is unable to produce cholesterol by its normal synthetic route, it will activate systems to scavenge cholesterol from various deposits. The liver will make apolipoproteins, such as Apo A-I, which is a precursor for HDL (see appendix D for descriptions of apolipoproteins and lipoproteins). Consequently, HDL levels will generally increase during statin therapy. These apolipoproteins can scavenge cholesterol from atherosclerotic fatty streaks if they have not yet been calcified (see chapter 6). This can help to reverse cholesterol deposition in the lining of arteries. The decrease in LDL-cholesterol and increase in HDL-cholesterol results in a favorable blood lipid profile.

Statins have become the most widely prescribed drugs in the world, with more than one hundred million people having taken them at some time and more than $50 billion in annual sales. Their promoters claim they are the new wonder drugs. They lower blood cholesterol levels, with dramatic effects on LDL-cholesterol. Many researchers publicizing their clinical studies claim decreases of as much as 30 percent in LDL-cholesterol levels with statin therapy, and similar decreases in cardiovascular events as a result (Gould et al. 2007). Many of the reports look quite impressive and have led many physicians and researchers to push for invariably more aggressive use of statins to combat heart disease. However, it is necessary to look a little more closely at the data to see how good these drugs really are.

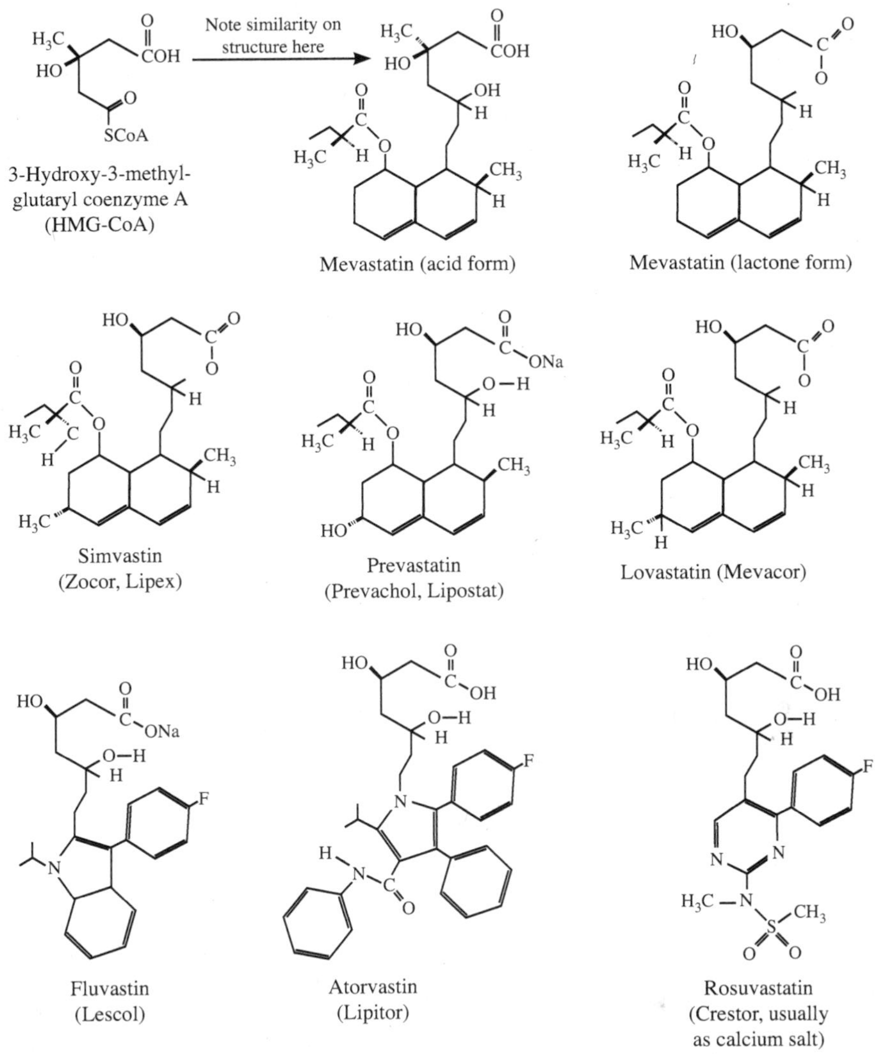

FIGURE 8.1. The chemical structures of 3-hydroxy-3-methyl-glutaryl coenzyme A (HMG-CoA) and several common statin drugs for comparison. Note the similarity in the glutaryl group, either as lactone, acid, or sodium salt, at the top of each statin structure.

Early Clinical Trials Show Promise

One of the first randomized, placebo-controlled clinical trials of statins, known as the 4S-Study, was published in 1994 (Scandinavian Simvastatin Survival Study 1994). The results were most impressive, with relatively few adverse effects from drug treatment. All 4,444 patients in the study had existing heart disease (previous myocardial infarction or angina pectoris) and relatively high LDL-cholesterol levels. There was a 35 percent decrease in LDL-cholesterol with statin therapy, and there were 30 percent fewer deaths among statin-receiving patients compared with the placebo-taking control group (8 percent versus 12 percent died). The number of coronary events was also much lower in the statin-treated group compared to the control group (19 percent versus 28 percent).

Another early randomized, placebo-controlled study was the West of Scotland Coronary Prevention Study (WOSCOPS) published in 1995 (Shepherd et al. 1995). This study showed similar results: 26 percent reduction in LDL-cholesterol, 31 percent reduction in cardiovascular events, 28 percent reduction in definite cardiovascular deaths, and 22 percent reduction in death from any cause (which was borderline in terms of being statistically significant, with $p = 0.051$) relative to the placebo-taking control group. The researchers indicated that the reduction in LDL-cholesterol alone did not account for the clinical benefit of statin therapy (WOSCOPS 1998). The statin drugs also reduce inflammation and decrease levels of C-reactive protein (CRP), another risk factor discussed in the previous chapter. The finding of a greater-than-expected decrease in cardiac events among statin-treated versus placebo-taking subjects was based on comparing results for subjects in each group with similar LDL levels. Such comparisons are rarely done. The statins were doing much more than merely lowering total serum cholesterol or LDL-cholesterol (see below).

A third study in Australia and New Zealand looked at the effects of prava-statin on cardiovascular events and death in men with previous heart disease and a broad range of baseline serum cholesterol levels (LIPID 1998). Again the results were similar: 29 percent reduction in all cardiovascular events, 24 percent reduction in cardiovascular deaths, and 22 percent reduction in deaths from all causes. The changes in serum cholesterol levels were similar, although changes in each class of lipoproteins were divided into three subgroups (low, intermediate, and high) according to baseline values. The greatest change in LDL-cholesterol in response to statin treatment was in the subgroup with highest levels at the start. Each of these reports indicated the drug was well tolerated, with relatively few adverse effects from treatment.

After several placebo-controlled studies of statins showed a clear benefit of drug treatment relative to cardiovascular events and death from heart disease, use of placebo in further studies was considered unjustified. Many subsequent trials then compared different dosage levels of a drug, or one statin drug versus another. There has been much discussion about subsequent studies that have given less convincing results with statin therapy. Often reports tend to downplay adverse side effects or dismiss them as rare or few. There seems to be a trend now toward more aggressive treatment with statins to reach a minimum level for LDL-cholesterol. But doubling the dosage does not mean

LDL-cholesterol will decrease by twice as much. Yet increasing dosage will surely increase the number and severity of adverse effects of these drugs, which are not insignificant.

Statins Are Not Just Lipid-Lowering Drugs

In addition to lowering cholesterol, statins also decrease levels of C-reactive protein (CRP), a powerful marker for cardiovascular disease risk (Albert et al. 2001). The decrease in CRP in response to statin therapy indicates a reduction in inflammation. Inflammation in the vascular system is usually considered to be what triggers the disruption of the fibrous cap over atherosclerotic lesions, which can lead to clotting, which in turn causes a myocardial infarction. In addition, there is evidence that statins can suppress atrial fibrillation and other types of cardiac arrhythmia (Young-Xu et al. 2003), which are frequently a cause of heart attacks. The statins also improve endothelial function, vascular tone, and blood flow (Paoletti, Bolego, and Cignarella 2005).

A 2005 review article identified twenty-two clinical trials of the pleiotropic (nonlipid) effects of statin drugs (Davignon and Leiter 2005). Ten of the studies in the review assessed (or were still assessing) various indicators of inflammation, such as CRP, and associated effects on plaque stability. Some of the studies assessed effects of statins on clotting factors, hypertension, levels of oxidized LDL, kidney disease, stroke, osteoporosis, Alzheimer's disease, and multiple sclerosis. Some studies were looking at multiple factors. There is much speculation that statins may be useful therapeutic drugs for many diseases, although the evidence for diseases other than cardiovascular disease is less compelling. Often the suggestion for usefulness in treating other diseases has been due to an observed secondary effect in clinical trials designed to study efficacy of statins in cardiovascular outcomes. For example, there appeared to be a reduced risk of depression in cardiac patients receiving statin therapy, although reduced levels of cholesterol in some studies have been associated with increased incidence of depression and suicide (Shafiq et al. 2005). There have also been reports of decreased incidence of Alzheimer's disease among statin users and lower risk of dementia among users, even though other studies have shown that low serum cholesterol has been associated with cognitive decline in aging populations (Shafiq et al. 2005).

In general, a decrease in risk factors for heart disease also decreases the risk for stroke. Statins can decrease the overall incidence of stroke, with about a 15 percent decrease in strokes caused by clotting (thromboembolism) but a 19 percent increase in strokes caused by hemorrhage has also been reported (Law, Wald, and Rudnicka 2003). The majority of strokes in the United States are caused by blood clots rather than hemorrhage. In addition, although the risk of nonfatal strokes was reduced with statin therapy, the risk of fatal strokes was not changed significantly (Cheung et al. 2004).

Explanations for Adverse Effects of Statins

In addition to lowering cholesterol, the statins also reduce levels of some other lipids in the body that require HMG-CoA reductase for their synthesis. One of these lipids is farnesyl pyrophosphate, which is involved in cell signaling and cell proliferation. There has been speculation that statins should suppress the proliferation of cancerous cells, and many studies of cells in culture have shown this to be true. However, numerous analyses of outcomes from the effect of statins on cardiovascular disease have indicated that statins neither promote nor suppress most cancers. There will be more on statins and cancer in chapter 10.

Another lipid requiring HMG-CoA reductase for its synthesis is coenzyme Q_{10} (CoQ, ubiquinone or ubiquinol), which is a component of the mitochondrial respiratory electron transport chain. There is concern that lower levels of this coenzyme could interfere with respiration and energy metabolism. There have been conflicting reports regarding the levels of coenzyme Q in patients receiving statins. Decreased coenzyme Q production would affect respiratory ATP production in mitochondria. This could affect ion transporters in the muscles, brain, and other organs, giving rise to the main side effects of these drugs, which are various degrees of pain and myopathy (muscle disease) in muscles and cognitive deficits.

A small study comparing coenzyme Q and vitamin E supplementation in patients taking statins indicated a significant decrease in muscle pain resulting from statin therapy when coenzyme Q supplements were taken but no change when vitamin E was taken. Vitamin E and coenzyme Q are two important lipid-soluble antioxidants in the body. As a component of the mitochondrial respiratory chain, coenzyme Q undergoes reversible oxidation and reduction. It acts as a shuttle for electrons between complex I (NADH dehydrogenase) or complex II (succinate dehydrogenase) and complex III (cytochrome b/c_1 complex or ubiquinol-cytochrome c oxidoreductase). It seems to be loosely bound to the inner mitochondrial membrane and free to diffuse within the membrane, rather than being firmly bound to membrane proteins like many of the other components of the respiratory chain. If it interacts with a stray oxygen molecule rather than its normal electron acceptor, complex III, it may generate a free radical. In this capacity, it would become a pro-oxidant rather than an antioxidant (Sohal and Forster 2007). It seems that free radical generation is directly correlated with coenzyme Q_9, a shorter version of coenzyme Q that may come from dietary sources, and is inversely correlated with coenzyme Q_{10} (Sohal et al. 2006).

Since coenzyme Q plays such a vital role in energy metabolism and statins interfere with its endogenous synthesis, coenzyme Q dietary supplements would seem to be a sensible suggestion for patients on statin therapy. Although coenzyme Q_{10} is routinely prescribed to patients on statin therapy in Europe and Canada, the medical establishment and pharmaceutical industry in the United States have not made this a standard recommendation. Perhaps such a recommendation would be admiting that the drugs have potential to cause adverse effects, whereas side effects have been downplayed all along.

It appears there has been only one short-term (four weeks) toxicity study of coenzyme Q supplements: it was done in Japan (Hosoe et al. 2007). That study showed that plasma levels of the coenzyme increased with dosage up to 300 mg per day with no noticeable adverse effects. Consumption of moderate doses of this supplement during statin therapy seems warranted. However, patients should be advised that this coenzyme can also act as a pro-oxidant that generates free radicals, and excessive doses could be harmful. There is little known about whether coenzyme Q from dietary supplements finds its way into the mitochondrial membrane. One study in China showed no significant change in muscle concentration of coenzyme Q with 150 mg per day supplements in healthy volunteers, and there was no significant change in several exercise performance parameters. It would be valuable to have a study in which levels of coenzyme Q are measured in mitochondria of statin-treated patients taking coenzyme Q supplements versus patients taking a placebo. This can be done easily by taking muscle biopsies. Studies could be designed whereby patients do not receive supplements for a period of time after statin therapy starts and then receive supplements.

Statins can cause minor aches in muscles (myalgia) or more severe myopathy that is characterized by increased levels of creatine kinase (a muscle enzyme) in blood (serum). Elevated serum levels of this protein are indicative of damage to muscle tissue. In some cases creatine kinase may be elevated by as much as one hundred times normal levels, indicating extensive muscle necrosis (rhabdomyolysis). Myopathies are reported in only about 0.1 percent of patients taking statin drugs, although there is some concern over whether there may be many overlooked or underreported cases in the aged population (Silva et al. 2007). If elevated creatine kinase is diagnosed in early stages, the effects may be reversible, although repair of muscle tissue is much slower in the elderly population. The alarm is not usually triggered until creatine kinase levels reach more than ten times what is considered a high normal level.

Another common side effect of statin drugs (in about 1 to 1.5 percent of patients) is elevated levels of serum transaminase (alanine aminotransferase), which is characteristic of liver disease. Administration of statin drugs usually results in about a threefold increase in levels of serum transaminase above the upper limit of normal, which is not surprising since many drugs can produce serum elevations in this liver enzyme. This is usually referred to as asymptomatic serum transaminase elevation, since there are usually no overt symptoms of liver disease displayed in patients. Symptomatic toxicity has been reported with all the statins, with symptoms including jaundice, hepatitis, cholestasis, fatty liver, cirrhosis, and acute liver failure, but these are considered rare (Corsini 2003).

The liver is responsible for metabolizing most drugs, including the statins, for elimination from the body. Often drugs will compete for cytochrome P450 enzymes that metabolize many drugs, so when multiple drugs are being consumed simultaneously, there may be interactions of these drugs via competition for the enzymes responsible for their metabolism. This results in increased levels of drugs in the body and greater likelihood of undesirable effects. With elderly and high-risk populations taking complex combinations of drugs, it can be next to impossible to determine all the possible drug and nutrient interactions.

As with any medication, the risks and benefits of statins must be evaluated to determine whether a patient's condition makes it worthwhile to pursue this type of drug therapy. Pharmaceutical companies have been given free reign to advertise these so-called wonder drugs directly to the public in the United States. Statins certainly have been shown to be effective in reducing cardiovascular events, but how efficacious are they at reducing overall death or improving quality of life for those with moderate or low risk? Many studies only report the decrease in cardiovascular events but neglect to report overall survival. To be fair, most of the studies performed in recent years compared one statin with another or with different doses of a statin drug, since early studies showed the efficacy of the drugs in lowering cholesterol and treating heart disease. It is curious that many patients (25 to 30 percent) discontinue their use of statins for various reasons, including economic factors (Peterson and McGhan 2005). The rate of continuation is better for statins than for several other types of lipid-lowering medications, such as bile acid sequestrants and fibrates, which commonly cause gastrointestinal disorders (Kamal-Bahl et al. 2007).

The point is that statins, although effective in lowering serum cholesterol (especially LDL-cholesterol) and reducing the incidence of cardiovascular events, may not be the wonder drugs with few or rare side effects that they are made out to be. More aggressive use of these drugs, with higher doses and treatment of larger segments of the population, may not be the best strategy for improving our collective general health. Statins may indeed be a magic bullet for people with verified heart disease or multiple risk factors, including diabetes, hypertension, and a family history of heart disease. For such high-risk individuals, the argument for statin therapy is certainly stronger. However, the emphasis on monitoring LDL-cholesterol during statin therapy and setting a goal of lowering LDL-cholesterol to as low as 70 mg/dL (by some advocates) seems to overlook the numerous pleiotropic effects of the statin drugs. The Adult Treatment Panel III of the National Cholesterol Education Program set a treatment goal of less than 100 mg/dL of LDL-cholesterol in high-risk patients (Grundy et al. 2004). Even this seems quite ambitious, since average LDL-cholesterol was 129 mg/dL in the United States when statin therapy was introduced and decreased to 123 mg/dL in 2002 with broad statin use (Carroll et al. 2005). The average level of cholesterol in the United States is considered to be too high by many advocates of lipid-lowering protocols.

Increasing life span as a result of fewer cardiovascular complications may include adverse side effects that diminish quality of life as part of the cost, particularly in the older population. As one group of critics pointed out, a nonfatal heart attack or stroke may be successfully managed, whereas some of the known adverse effects, such as rhabdomyolysis (severe muscle wasting), polyneuropathy (multiple peripheral nerve disorders), aggressive or suicidal behavior, amnesia, and serious congenital defects (birth defects among the children of mothers taking statins) can be catastrophic for patients (Ravnskov, Rosch, and Sutter 2005).

There is much controversy regarding underreporting the adverse effects of these drugs. Rhabdomyolysis is a severe form of muscle disease that is rarely seen as a statin-induced symptom (fewer than 0.1 percent of patients) (Davidson and Robinson 2007), but how prevalent is some loss of muscle function or some increase in muscular aches and pains? Cognitive problems (memory loss and confusion) may be dismissed by primary care physicians, especially in elderly patients on statin therapy, as just one of the consequences of aging. The possible connection between some peripheral neurologic problem and statin therapy may be overlooked and attributed to unknown etiology. Minor aggression may be considered just part of a person's personality, and aggravation of such a tendency may go unnoticed until it is too late in the most violent cases, or may appear some time after initiation of statin treatment, concealing any connection between drug treatment and change in behavior.

Conclusions

The statin drugs are enjoying enormous popularity in the cardiac health care community. They are highly effective drugs for improving the serum lipid profile, lowering LDL-cholesterol, raising HDL-cholesterol, and often lowering serum triglycerides. Their impact on reduction of cardiovascular events in the high-risk population is most impressive. Some studies did not find significant changes in death from all causes when statin therapy was compared with placebo controls, although other studies have shown an overall improvement in life span.

The possible therapeutic effects of statin drugs on several health conditions are being explored, in addition to their cardiovascular benefits. Their favorable effects on inflammation, clotting factors, atrial fibrillation, and other factors associated with cardiovascular disease may be playing as much of a role in reducing coronary events as their cholesterol-lowering properties. Perhaps it would be more practical to look at the numerous pleiotropic effects these drugs have when assessing their effectiveness in reducing heart disease risk, rather than focusing on the reduction in LDL-cholesterol. Setting treatment goals for LDL-cholesterol levels that are far below the national average means more aggressive prescription of these drugs. It is becoming clear that higher doses of these drugs and more general use of them are accompanied by increases in adverse side effects. We should not lose sight of quality of life considerations when considering approaches to better health, particularly in lower- or moderate-risk populations.

Diet and lifestyle modification, especially exercise, should be the first consideration of anyone who may feel at high risk for heart disease. The earlier one starts the better.

9

Inflammation, Anti-inflammatory Drugs, and Lipid Mediators

Willow bark was known since ancient times to reduce fever and decrease pain. An active extract of willow was isolated in the late eighteenth century and given the name salicin, from the Latin *Salix* for willow, or trees in the Salicaceae family. In the early nineteenth century salicylic acid was identified as the active component and was neutralized to form potassium or sodium salicylate salts. The French chemist Charles Gerhardt first converted salicylic acid to its acetyl ester in 1853, but this derivative seemed to be ignored for the next forty years. In 1896 Felix Hoffman, a chemist at the Bayer subsidiary of I. G. Farben in Germany, repeated Gerhardt's procedure and named the product aspirin. Heinrich Dreser reported its effectiveness for the treatment of fever and pain in 1899, which started a long reign for aspirin as the world's most widely used drug, a position of preeminence it held for nearly a century (Insel 1990).

Eventually, pharmaceutical companies designed other analgesic drugs, some working even better than aspirin for pain relief. However, aspirin still holds a special place in the pharmacopeia, not just as an analgesic, but as an anti-inflammatory, anticoagulant, and antipyretic (fever-reducing) drug. Second-generation analgesic and anti-inflammatory drugs captured a large share of the market since their introduction in the 1960s and 1970s, but they still have a long way to go to knock aspirin from the rank of largest-selling drug of all time in terms of overall quantity. The recent flurry over the latest anti-inflammatory drugs (COX-2 inhibitors) has shown that it will not be a trivial matter to replace aspirin.

As we saw in chapter 3, the mode of action for aspirin was worked out by John Vane and colleagues at the Wellcome Research Institute in Great Britain. They found that the acetyl group of aspirin is transferred to the active site of the enzyme cyclooxygenase (COX), inhibiting that enzyme from converting polyunsaturated fatty acids to bioactive substances known as prostaglandins and thromboxanes (Vane 1971). The acetyl group transferred is the one that Gerhardt and Hoffman attached to the salicylic acid to form aspirin (see figure 9.1). This acetyl group is attached to the COX enzyme permanently, making aspirin an irreversible inhibitor of COX. The acetyl group does not get removed from the enzyme. The only way the catalytic activity of the COX enzyme is restored to the cell is for the cell to make new enzyme via protein synthesis. This could take several hours in a typical cell but does not occur in blood platelets, because they do not have a nucleus with DNA, nor do they have the apparatus for synthesizing new proteins.

Aspirin is unique among the COX inhibitors in being irreversible. The other inhibitors are reversible, and enzyme activity returns when levels of the drug have dwindled owing to metabolic breakdown and excretion. This is why doctors often recommend a low dose of aspirin to people at high risk of heart attack. The low dose will be used up reacting with the first COX enzymes it encounters, primarily those in the blood, especially in blood platelets, which results in the observed anticoagulant effect with repeated low doses. A day or two after aspirin has been taken, a sufficient number of new platelets have been generated in bone marrow with active COX enzyme to make thromboxane and initiate blood clotting. It was recognized long ago that aspirin increases bleeding time by preventing blood clots (Roth and Majerus 1975). A person who has taken several aspirin tablets over a period of a few days will generally bleed

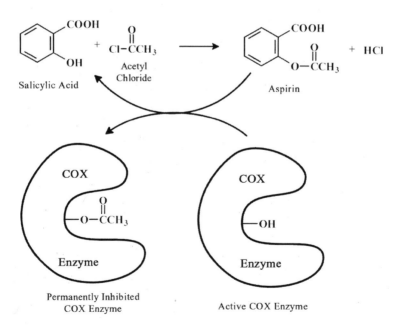

FIGURE 9.1. Conversion of salicylic acid to aspirin by chemical reaction with acetyl chloride (*top*). Aspirin reacts with cyclooxygenase (COX) by transferring the acetyl group to the enzyme, forming salicylic acid.

for a longer time from a cut or scratch as a result of aspirin's action on COX in the blood platelets.

Once the biochemical action of aspirin was understood, pharmaceutical companies put great effort into developing new drugs that would work like aspirin and could be patented. One drug developed in the 1960s was ibuprofen, which was sold as a prescription drug under the brand name Motrin. The U.S. Food and Drug Administration (FDA) approved the nonprescription sale of ibuprofen in 1984, when the patent expired, and it became widely available under the brand names Advil and Nuprin, as well as in generic form. Naproxen, another anti-inflammatory drug, came on the market under the brand name Aleve and also became a popular generic drug. The generic drug companies have enjoyed a large share of the market for anti-inflammatory drugs.

These drugs, along with aspirin, are collectively known as nonsteroidal anti-inflammatory drugs or simply NSAIDs. Some steroids, such as cortisone, are also anti-inflammatory, but rather than inhibiting the COX enzyme, they inhibit the enzyme that releases arachidonic acid from phospholipids in cell membranes (see figure 9.2). This latter enzyme is known as phospholipase A_2 (PLA_2). The anti-inflammatory steroids have far more side effects than NSAIDs because the steroids are also hormones that send signals throughout the body causing cells to alter their metabolism. The hormone actions do not necessarily involve prostaglandins and thromboxanes.

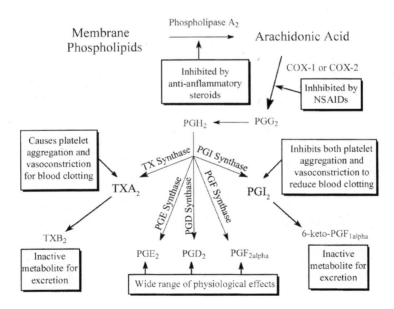

Figure 9.2. Synthesis of prostaglandins and thromboxanes from arachidonic acid.
Adapted from Clària and Romano 2005.

Aspirin has been the drug of choice to prevent a heart attack in patients at high risk because of its anticoagulant properties (Ames and Gold 1991). The platelets are responsible for initiating a blood clot by forming TXA_2, which causes platelets to stick together and blood vessels to contract when a blood vessel is severed. However, in the case of a heart attack, the blood vessels need not be severed to trigger the events that lead to a blood clot (as described in chapter 6). This is often triggered by an immune response to bacterial infection (such as *Chlamydia pneumoniae*) and the resulting inflammatory reaction in vascular tissue (Song et al. 2000).

If COX in the platelets has been irreversibly inhibited by aspirin, the blood clot may not get started or may remain small, allowing a ruptured cap to be repaired without a life-threatening cardiovascular event. If a patient at risk of myocardial infarction has been consuming, in addition to aspirin, an abundance of omega-3 fish oils relative to omega-6 vegetable oils, the clotting response in the above scenario would be further diminished. Eicosapentaenoic acid (EPA) derived from fish oils also suppresses inflammatory reactions in the vascular system, resulting in less damage to the fibrous cap and less chance of myocardial infarction.

There is concern about increased incidence of stroke among patients taking aspirin regularly, which can lead to cerebral hemorrhaging (due to decreased blood clotting) (Kronmal et al. 1998). A high intake of omega-3 fatty acids from eating a lot of fish or regular consumption of fish oil supplements, especially EPA, could augment the anticlotting effect of aspirin and make an individual even more prone to hemorrhaging. Therefore daily low doses of aspirin may not be a panacea for the general population.

COX-2 Inhibitors Arrive on the Scene

A newer generation of drugs that block the action of COX came on the market around the turn of the twenty-first century—a century after aspirin appeared. These are known as COX-2 inhibitors because they are relatively selective for the COX-2 form of the enzyme that produces prostaglandins from arachidonic acid and have little effect on COX-1. Traditional NSAIDs inhibit both forms of the enzyme, although they may exhibit subtle differences. For example, aspirin and ibuprofen have several-fold greater preference for COX-1 versus COX-2 (Mitchell et al. 1993). There are a few different forms of COX in the body, which are distinguished by numbers, namely COX-1, COX-2, and COX-3, with COX-1 and COX-2 being more abundant and better understood.

COX-1 is considered a constitutive enzyme, meaning that it is present all the time in tissues where it is found, such as in the gastric mucosa (stomach lining) and blood platelets. COX-2 is an inducible form of the enzyme, meaning that it is usually present at low levels, but some event can trigger cells to produce it in larger quantities. Inflammation resulting from injury or infection is known to induce certain cells to produce this enzyme. Both forms of the COX enzyme catalyze the same reaction, namely conversion of arachidonic acid to prostaglandin G_2, followed by a subsequent peroxidase reaction to form prostaglandin H_2 (see appendix C for structures). Other enzymes catalyze formation of different end products from PGH_2 in different cells, such as thromboxane synthase to form TXA_2 in platelets and prostacyclin synthase to produce PGI_2 (prostacyclin) in vascular endothelial cells (Clària and Romano 2005) (see figure 9.2).

Once different forms of the COX enzyme were identified, pharmaceutical companies were eager to capitalize on their differences. It was recognized that immune cells involved in the inflammatory response make predominantly COX-2. An effort was made to develop drugs that would specifically inhibit COX-2 and spare COX-1. The latter is present in certain tissues all the time and associated with many physiological activities, such as protecting the gastric mucosa and initiating blood clots when we are bleeding. It was believed that selective COX-2 inhibitors should cause fewer gastrointestinal problems than nonselective COX inhibitors (NSAIDs).

In 1999 celecoxib (sold by Pfizer under the brand name Celebrex) became the first COX- 2 inhibitor to reach the market. Rofecoxib (sold by Merck under the brand name Vioxx) immediately followed. Although there were not yet published reports regarding lower incidence of gastrointestinal disturbances, bold marketing resulted in celecoxib setting a record for sales of a pharmaceutical in the shortest period of time—more than $3 billion worldwide in 2001 (Juni, Rutjes, and Dieppe 2002).

Gastrointestinal problems from aspirin and other NSAIDs affect only about 5 percent of the population. A comparison of rofecoxib (Vioxx) with naproxen showed that both drugs provided similar relief of joint stiffness and pain in rheumatoid arthritis patients. Rofecoxib produced about 2.1 gastrointestinal events (perforation, obstruction, bleeding, and symptomatic ulcers) while naproxen caused about 4.5 such events per 100 patient-years (per 100 users of the drugs over a period of one year) (Bombardier et al. 2000). Naproxen shows a preference for COX-1 over COX-2 similar to that of aspirin and ibuprofen.

Two large-scale clinical trials of the COX-2 inhibitors were initiated in 2000: the Celecoxib Long-Term Arthritis Safety Study (CLASS) and the Vioxx Gastrointestinal Outcomes Research (VIGOR) trial. These randomized, placebo-controlled studies compared celecoxib and rofecoxib with nonselective NSAIDs, namely with ibuprofen and diclofenac in the CLASS trial and with naproxen in the VIGOR trial. Early reports showed significantly lower incidence of complicated ulcers with the COX-2 inhibitors; however, there was a higher incidence of serious adverse events with the COX-2 inhibitors, including deaths, admissions to hospital, and various life-threatening or serious disabilities (Wright 2002). Overall mortality with COX-2 inhibitors was greater than with traditional NSAIDs, but the differences were not statistically significant. Eventually, when all the data were released, the VIGOR study indicated a significant increase in cardiovascular events with rofecoxib use compared with naproxen.

There was some concern soon after the COX-2 inhibitors were introduced that they might cause heart attacks because of their selectivity for COX-2 versus COX-1. In 2002 Nobel laureate John Vane explained how COX-2 activity in vascular endothelial cells leads to production of prostacyclin (PGI$_2$), a prostaglandin that causes vasodilation and inhibits platelet aggregation. These actions offset the prothrombotic actions of TXA$_2$ produced via COX-1 in the platelets. In other words, COX-2 inhibitors tipped the balance between clotting and anticlotting factors to strongly favor clotting, whereas nonselective NSAIDs inhibit both COX enzymes and suppress both clotting and anticlotting activities (see figure 9.3). The concern was realized when the VIGOR trial showed that patients treated with Vioxx experienced a fivefold higher incidence of heart attacks compared with patients treated with naproxen.

The primary aim of the large clinical trials was to establish their safety with regard to gastrointestinal problems. The NSAIDs chosen for comparison may have resulted in celecoxib not producing a statistically significant increase in heart attacks. Diclofenac has a preference for COX-2 over COX-1, and ibuprofen has less preference for COX-1 over COX-2 than naproxen (Warner et al. 1999). Vioxx remained on the market for two years after the problem surfaced but was eventually withdrawn, along with Bextra (valdecoxib) some months later. Celebrex remains on the market, but the FDA requires a "black box" warning on the label, which is the most serious warning a medication can carry.

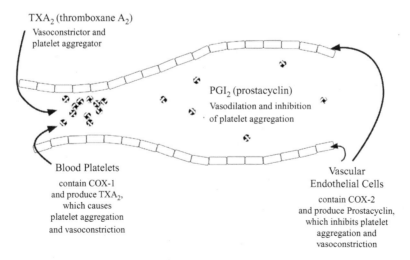

FIGURE 9.3. Most NSAIDs inhibit both COX-1 and COX-2, inhibiting both platelet production of TXA$_2$ and vascular production of PGI$_2$. COX-2 inhibitors do not block TXA$_2$ production by platelets (on the left) but inhibit vascular production of PGI$_2$ (on the right), and consequently they promote blood clotting.

New Actions for an Old Wonder Drug

At about the time the new COX-2 inhibitors were coming to market, scientists were discovering new ways in which aspirin works its wonders. Aspirin modifies the COX-2 enzyme in neutrophils (immune cells involved in the inflammatory response) by transferring its acetyl group to the active site of the enzyme (see above) (Clària and Serhan 1995). Instead of completely shutting down the enzyme, aspirin modification causes COX-2 to catalyze a different chemical reaction. Instead of forming prostaglandins G_2 and H_2 from arachidonic acid, a substance known as 15-HETE (15-hydroxyeicosatetraenoic acid) is produced (see figure 9.4). This lipoxin is normally formed when a different enzyme known as 15-lipoxygenase catalyzes the reaction, but aspirin-modified COX-2 forms a different isomer (or epimer) of 15-HETE.

The aspirin-modified COX-2 also converts eicosapentaenoic acid (EPA) and docosahexaenoic acid (DHA) to hydroxylated products that have potent anti-inflammatory activity. This is a new and important role for these omega-3 polyunsaturated fatty acids from fish oils as a result of this unique action of aspirin with COX-2. Aspirin seems to completely block all enzyme activity of COX-1. Although these two enzymes normally catalyze the same reaction, formation of PGH_2 from arachidonic acid, they are different proteins derived from different genes and have discrete structures and reactions to aspirin therapy.

Lipoxins are known to be synthesized by an unusual metabolic pathway involving enzymes from more than one type of cell. Because they are lipids, they can pass through membranes quite easily and may travel from one cell to another. Enzymes in other cells add more oxygen atoms (more specifically hydroxyl groups) to 15-HETE, until eventually it has three hydroxyl groups. These substances are normally formed in the body, but aspirin results in formation of a right-handed (rectus in chemical terminology) structure rather than the left-handed (sinister) structure that 15-lipoxygenase forms. The right-handed form of lipoxin has as much or more biological activity compared with the left-handed form of this potent bioactive product. Since COX-2 is in greater abundance when there is inflammation, the aspirin-triggered alteration of this enzyme now supplements the activity of any 15-lipoxygenase that may be present.

Curiously, aspirin seems to be the only drug that causes transformation of COX-2 to a different form of enzyme because of its unique transfer of the acetyl group. Lipoxins have potent anti-inflammatory activity, decrease myocardial infarctions, and inhibit cell proliferation in colorectal adenomas (cancers). They have specific receptors in certain types of cells that inhibit release of several other proinflammatory signals, such as interleukin-1 beta (IL-1β), tumor necrosis factor-alpha (TNF-α), and interleukin-8 (IL-8), while stimulating release of a more favorable messenger, interleukin-4 (IL-4) (Serhan et al. 2002).

For decades it was believed that omega-3 oils suppress inflammation by inhibiting formation of potent proinflammatory agents formed from omega-6 fatty acids. Many of the omega-3 eicosanoid products are known to be less active in producing an inflammatory response than omega-6 derivatives from arachidonic acid (derived from vegetable oils). With the discovery of aspirin-modified activity of COX-2, there is now another mechanism by which omega-3

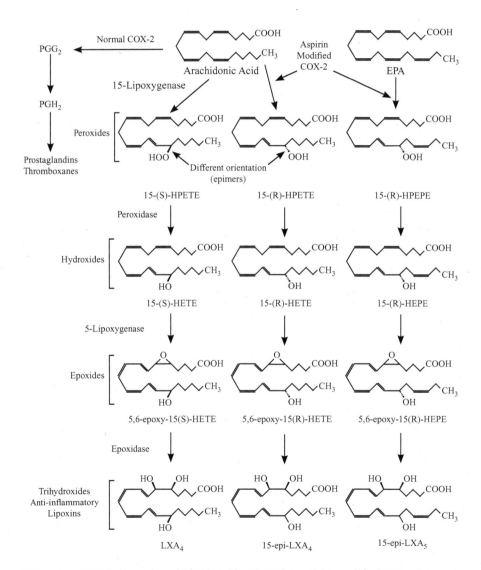

Figure 9.4. Metabolism of arachidonic acid and EPA by aspirin-modified COX-2 followed by metabolism by 5-lipoxygenase to produce the corresponding 15-*epi*-lipoxins, which are anti-inflammatory agents. Normal COX-2 metabolism of arachidonic acid to form prostaglandins is shown on the left for comparison.

fatty acids, including DHA, can suppress inflammation. The effect can occur without aspirin administration but may be more effective when aspirin is taken. Clinical trials to demonstrate the efficacy of combined omega-3 supplementation with aspirin have not been done, but this would seem to be a worthwhile study. However, this combination of anticlotting agents could promote hemorrhaging, which can be as life-threatening as too much clotting.

Overview of Other Inflammatory Syndromes

Aurelius Cornelius, a Roman physician (30 B.C.E. to 45 C.E.), described inflammation as rubor (redness), calor (heat), tumor (swelling), and dolor (pain) (Yoon and Baek 2005). Prostaglandin E_2 (PGE_2) is believed to play a major role in these symptoms by increasing blood flow through tissues as a result of its vasodilator action (see figure 9.5). It also increases vascular permeability, which leads to fluid and other components leaving the vascular system to cause edema (swelling) and redness. Swelling and pressure causes the sensory nerves in the tissue to send pain signals to the spinal chord and brain (Clària and Romano 2005). PGE_2 increases the sensitivity of sensory nerves and acts as a mediator to induce fever.

Local production of heat occurs in the inflamed area(s) as a result of increased metabolic activity of white blood cells, without a systemic temperature increase. Increased vascular permeability allows leukocytes to infiltrate the area in response to another eicosanoid, leukotriene B_4 (LTB_4). There is a wide array of signaling factors produced in the inflamed tissue, including histamine and many cytokines (interleukins, interferons, and tumor necrosis factor-alpha).

Most diseases discussed here are autoimmune diseases, whereby the body's immune system attacks certain tissues. It is believed that certain types of infectious agents (viral, bacterial, and fungal) trigger an immune response that may

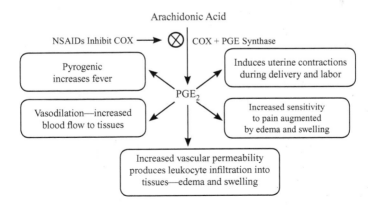

FIGURE 9.5. Physiological actions of prostaglandin E_2.

evoke an errant attack on oneself (Sherwood 2004, 441). Substances in the body's own tissues may resemble antigenic (immune system–activating) material on the infectious agent. Sensitized immune cells are unable to distinguish self from invasive organisms. This was mentioned in chapter 6 regarding *Chlamydia* infections stimulating the immune system to attack the protein cap over atherosclerotic lesions to initiate a blood clot in the heart. The sophisticated intricacies of the immune system manage to work well most of the time, but we notice errant responses and ignore the effective ones.

Inflammation is an underlying cause of a wide range of afflictions, ranging from arthritis, asthma, and allergies to eczema, heart disease, psoriasis, lupus (systemic lupus erythematosus), multiple sclerosis, migraine headache, and ulcerative colitis, to name a few. The *Merck Manual* indicates there are more than one hundred specific conditions related to inflammation and rheumatic disorders (Berkow 1982, 1166). Why these problems flare up in some individuals and not in others who may be doing the same things is not known. As with susceptibility to most diseases, genetics plays a major role. Eicosanoids derived from polyunsaturated fatty acids are involved, and dietary fats and oils are likely factors in determining severity of symptoms.

Numerous studies have assessed the effects of dietary omega-3 supplements on inflammation and autoimmune diseases, particularly rheumatoid arthritis (Simopoulos 2002b). In addition to omega-3 oils, other dietary fats that affect eicosanoid metabolism have been tried. Dietary saturated fats (from beef tallow) dramatically reduced adjuvant-induced arthritis in rats compared with proinflammatory omega-6 vegetable oils (Lawrence 1990). In general, supplements that suppress arachidonic acid–derived eicosanoids have favorable effects. In some clinical trials of omega-3 supplements, no significant beneficial effects were observed with dietary supplements alone, but often there was a reduced requirement for NSAIDs to reduce pain and inflammation. Patients should be advised to decrease their consumption of omega-6 vegetable oils while taking omega-3 supplements in order to experience optimal effects with omega-3 supplements (Cleland and James 1997).

Standard therapy for arthritis is NSAIDs, with recourse to corticosteroids such as cortisone or hydrocortisone for cases that do not respond well to NSAIDs. The steroids control inflammation by inhibiting release of arachidonic acid from membrane phospholipids. Anti-inflammatory steroids suppress production of all eicosanoid metabolites, not just prostaglandins and thromboxanes. Leukotriene B_4 is a chemotactic agent derived from arachidonic acid that attracts white blood cells to the inflamed area and augments inflammatory symptoms. NSAIDs do not suppress leukotriene production, which may explain why some patients do not respond well to NSAID therapy for arthritis and other inflammatory autoimmune diseases. On the other hand, PGE_2 is involved (as described above), and its production is inhibited by NSAIDs. A wide range of NSAIDs, including many prescription and nonprescription drugs, are used to treat arthritis, with a range of efficacy in different individuals and varying undesirable side effects.

Inhibition of COX-2, whether by selective COX-2 inhibitors or by NSAIDs, results in impaired healing of broken bones (Simon, Manigrasso, and O'Connor 2002) and increases deterioration of joints in patients with osteoarthritis of the knee (Huskisson et al. 1995). It is still unclear whether the conversion of COX-2 by the acetyl group from aspirin would allow alternative products to be formed that could stimulate the healing process in bone repair and osteoarthritis, but this effect may require greater levels of omega-3 fatty acids through supplements. Aspirin and other NSAIDs reduce the risk of colon cancer (Garcia-Rodriguez and Huerta-Alvarez 2001) and prostate cancer (Garcia Rodriguez and Gonzalez-Perez 2004). There are numerous reports of conflicting or contradictory results for one NSAID versus another. NSAIDs inhibit COX-1 and COX-2 to varying degrees (Warner et al. 1999), but expected differences with regard to therapeutic outcomes are not always observed. There may be other actions of these NSAIDs that are not yet understood, or there may be differential effects in different tissues and diverse responses due to genetic variation.

The effects of dietary omega-3 supplements on inflammatory bowel disease (Crohn's disease and ulcerative colitis) have been studied by several groups. In one placebo-controlled study, fish oil supplements resulted in decreased LTB_4 (from arachidonic acid) production and increased LTB_5 (from EPA), with an overall decrease in leukotrienes. Although clinical improvement and sigmoidoscopy results were considered modest, it was possible to reduce levels of anti-inflammatory steroid dosage to manage discomforts caused by the disease (Stenson et al. 1992). Several studies have shown positive results for fish oil supplementation in inflammatory bowel disease, although some have shown little or no improvement (Endres, Lorenz, and Loeschke 1999). Differences in response may reflect the heterogeneity of these diseases, as well as differences in experimental design in terms of limiting dietary intake of omega-6 fatty acids. Genetic factors influence levels of specific enzymes involved in eicosanoid metabolism, which has an impact on response to changes in the ratio of omega-6 to omega-3 fatty acids in tissues as a result of dietary manipulations.

Potent physiological effects may occur when arachidonic acid is abundant in body tissues as a result of consuming diets containing large amounts of omega-6 vegetable oils. Dietary fish oils and omega-3 supplements, by inhibiting arachidonic acid metabolism to bioactive eicosanoids and forming the omega-3 eicosanoid derivatives, influence the course of several inflammatory syndromes. Clinical studies have not always produced clear and consistent results. The relative availability of omega-3 versus omega-6 fatty acids in membranes can have a profound effect.

Artemis Simopoulos, president of the Center for Genetics, Nutrition and Health, compiled data on the ratio of omega-6 to omega-3 fatty acids in modern Western diets and argues strongly that diet is exacerbating many inflammatory diseases (Simopoulos 1991, 2002a). Although dietary changes may not completely alleviate various inflammatory diseases, changes in the omega-6 to omega-3 fatty acid ratio in the diet could make the symptoms much more tolerable and decrease or diminish reliance on anti-inflammatory drugs, especially powerful steroids that have a wide range of undesirable side effects. Small amounts of omega-3 EPA in a sea of omega-6 oils would have relatively little impact. Dietary changes that decrease omega-6 intake by elimination of vegetable oils may result in significant benefit, particularly when done in conjunction with omega-3 supplementation.

The Eicosanoid Connection in Asthma

Asthma is one of the most common chronic diseases of childhood and may afflict as much as 5 percent of that population in industrialized countries (Fleming and Crombie 1987). Bronchial asthma is characterized by increased numbers of inflammatory cells in the airway walls, which produce swelling of airway walls, increased secretion of mucous into the airways, and contraction of alveolar smooth muscles. The release of numerous chemical mediators, including leukotrienes and cytokines, can produce an allergic reaction. Leukot-rienes C_4, D_4, and E_4 are collectively known as slow-reacting substance (SRS) of anaphylaxis or sulfido leukotrienes because they are formed from the sulfur-containing peptide glutathione. These leukotrienes are derived from arachidonic acid and can increase vascular permeability in the lungs, stimu-late smooth-muscle contraction leading to vasoconstriction in blood vessels as well as bronchial constriction, and increase mucus secretion. An asthmatic response can be initiated by a variety of factors, including exercise, cold air, environmental pollutants, biological agents, and aspirin ingestion.

Fish oil (omega-3) supplements have been shown to improve pulmonary function in some asthmatics (responders) but not in others (nonresponders) (Broughton et al. 1997). When the relative level of omega-6 to omega-3 polyun-saturated fatty acids was high (ten to one), there was diminished respiratory capacity in asthmatics. On the other hand, a lower ratio (two to one) resulted in significant improvement in more than 40 percent of study participants following inhalation of methacholine (a drug that triggers an asthmatic response). Studies that showed little or no effect with fish oil supplementation did not report the levels of omega-6 fatty acids that were consumed (Broughton et al. 1997). Another study in Japan showed significant increases in plasma EPA and beneficial effects of omega-3 oil supplements in children with bronchial asthma in a strictly controlled environment (Nagakura et al. 2000).

Because leukotrienes play a major role in bringing on the symptoms of asthma, and leukotrienes formed from omega-6 vegetable oils are the most potent substances known in this capacity, it would seem that the large increase in reported asthma in industrialized nations may be due to high dietary intake of omega-6 vegetable oils relative to omega-3 oils. Many argue that increased incidence of asthma, particularly in cities, is due to air pollution, although there has been major progress in cleaning up polluted air in the past quarter century—a period that coincides with increased asthma. The perception that the number of asthma cases has been increasing may be due to increased severity of asthma symptoms, which would cause more patients to seek medical treatment. If asthma cases were milder in years past, they may not have prompted the patient to seek medical assistance and therefore went unnoticed.

Another factor that has been receiving much attention with regard to asthma is indoor air pollutants. In particular, formaldehyde (at 100 micrograms per cubic meter for 30 minutes) augments the allergic response to dust mite allergen in asthmatic patients (Casset et al. 2006). Chad Thompson and Roland Grafström at the U.S. Environmental Protection Agency have shown formaldehyde interferes with metabolism of a natural bronchodilator known as S-nitrosoglutathione, resulting in exacerbation of asthma by formaldehyde (Thompson and Grafström 2008). Formaldehyde is formed from fructose degradation under conditions that prevail in many sodas (Lawrence, Mavi, and Meral 2008). It would be important to determine whether formaldehyde from fructose degradation in soft drinks containing high-fructose corn syrup may be a significant factor in the increased incidence of asthma in children in the United States. This would be less of a factor in countries where high-fructose corn syrup is not commonly used. Sucrose does not decompose as readily as fructose, and drinks sweetened with sucrose would probably not have formaldehyde in them.

Some asthmatics have a condition known as aspirin-intolerant asthma. It is more common in adults than children. If aspirin-intolerant asthmatics take aspirin or other NSAIDs, they often develop an asthmatic attack within a couple of hours. Aspirin-intolerant asthmatics were found to produce five times as much enzyme for production of leukotriene C_4, LTC_4 synthase, as aspirin-tolerant asthmatics. The level of this enzyme was eighteen times higher in aspirin-intolerant asthmatics than in normal (nonasthmatic) individuals. Levels of other enzymes, such as COX-1 and COX-2, were not affected (Cowburn et al. 1998). It is believed that inhibiting the COX enzymes with aspirin will shunt more arachidonic acid (released in the initial activation of these cells) toward the lipoxygenase enzymes that lead to leukotriene production (figure 9.6).

Anti-inflammatory steroids are effective in aspirin-intolerant asthmatics because these drugs inhibit release of arachidonic acid from membrane phospholipids. Higher levels of lipoxin A_4, an anti-inflammatory eicosanoid, are also produced in aspirin-tolerant asthmatics compared with aspirin-intolerant asthmatics or normal subjects (Sanak et al. 2000). It was suggested that decreased lipoxin production may account for more severe symptoms in the aspirin-intolerant asthmatics. Perhaps a genetic defect in the enzyme that normally produces lipoxins is causing this diminished production, although such a genetic mutation has not been reported.

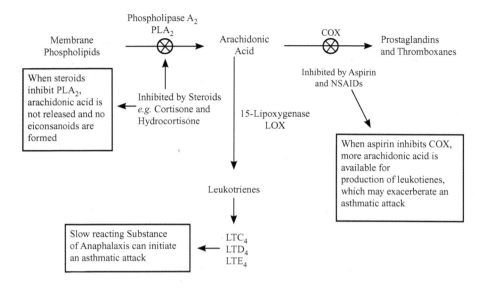

FIGURE 9.6. Steroids inhibit release of arachidonic acid from membrane phospholipids and suppress formation of all eicosanoids, whereas aspirin (and NSAIDs) inhibit COX and suppress formation of prostaglandins and thromboxanes, but COX inhibition can make more arachidonic acid available for leukotriene production, which may exacerbate asthmatic symptoms.

Inflammation and Heart Disease

It was mentioned in chapter 7 that C-reactive protein (CRP) is a better indicator for risk of heart attack than serum cholesterol. CRP levels increase as a result of the inflammatory response, although CRP could be a messenger that augments the inflammatory response and triggers immune cells to become more active. It is generally believed that aspirin reduces risk for heart disease by "thinning the blood," meaning it blocks the formation of thromboxane A_2 by platelets, which diminishes platelet aggregation and vasoconstriction. As an anti-inflammatory drug, aspirin also suppresses inflammation, which may be the underlying cause of many heart attacks.

The cholesterol-lowering drugs known as statins (see chapter 8) also decrease levels of CRP, indicating an anti-inflammatory role for these drugs. Increased levels of CRP can also induce atrial fibrillation and promote its persistence (Young-Xu et al. 2003). Atrial fibrillation and arrhythmia are believed to be major contributors to myocardial infarction. Statins lower a host of proinflammatory cytokines and increase anti-inflammatory cytokines. The statins have a wide range of effects on several physiological systems.

Conclusions

The omega-3 fatty acids that are abundant in fish oils have favorable effects in many inflammatory syndromes, but their beneficial effects could be diluted if large amounts of omega-6 fatty acids are present in body tissues as a result of consuming relatively large quantities of vegetable oils. Dietary saturated fats lack activity with regard to inflammation and its consequences, so saturated fats in the diet do not promote inflammatory syndromes.

The most widely promoted monounsaturated oil, olive oil, does contain significant amounts of omega-6 polyunsaturated fatty acids, generally 10 to 15 percent, compared with greater than 50 percent found in most vegetable oils. Because of its inherent flavor and the higher price for olive oil, it is often used in smaller amounts than inexpensive vegetable oils. Canola oil has a favorable fatty acid profile with respect to inflammation, since it has less than 25 percent omega-6 fatty acids and about 10 percent omega-3 fatty acids. Palm oil also has only about 10 percent omega-6 fatty acids and contains mostly monounsaturated oleic acid and saturated palmitic acid in nearly equal proportions. Palm oil lacks omega-3 fatty acids, but the abundance of monounsaturated and saturated fatty acids in palm oil should lower its potential for promoting inflammation.

Mayonnaise is composed of mostly vegetable oil and is a major source of omega-6 fatty acids in the diet of many people. This is another food to be aware of when trying to reduce omega-6 oils in the diet in order to reap the greatest benefits from omega-3 fish oils. The agricultural industry has been developing seed plants with more favorable profiles of fatty acids in their oils, although decisions are often based on cholesterol-lowering effects. Food processors are modifying their formulations to be able to claim they are providing consumers with healthier alternatives, but their focus, too, is generally on lowering cholesterol. All polyunsaturated fatty acids will help to lower serum cholesterol, but people afflicted with pain from inflammation must be made aware of the adverse effects the omega-6 vegetable oils can have on inflammatory syndromes. A list of various oils and their fatty acid profiles is given in appendix B.

Finally, it seems that omega-3 fish oil supplements can provide some benefit in many inflammatory diseases. In addition, the combination of aspirin with the omega-3 supplements may be more effective than either of these therapies taken alone. Anyone contemplating such experimentation on a personal level should recognize that the combination of these two anticlotting agents (omega-3 oils and aspirin) could promote excessive internal bleeding that would be noticed as bruising or hematomas (large patches of black and blue, especially in deep tissue) and could lead to life-threatening internal hemorrhaging in more susceptible individuals.

10

Cancer and Immunity

The products of lipid peroxidation are many and can have a wide range of biochemical and physiological effects. Some of the products are known to promote cancer. Lipid peroxidation is a free-radical process, and the reactive oxidation products involved can spread to other venues to wreak havoc throughout the cell where lipid peroxidation is taking place. In some cases the havoc wrought can result in the cell coordinating its own demise, a process known as apoptosis or programmed cell death. Apoptosis can spread to neighboring cells, resulting in many cells dying in a localized area. The fact that damaged cells can sacrifice themselves suggests that apoptosis provides an evolutionary advantage to the greater organism. Before we consider the role of dietary fats in cancer, it may be helpful to review the biochemical processes involved in carcinogenesis.

Carcinogenesis

Early models of carcinogenesis invoked a two-stage mechanism to represent the major events that lead normal cells to become cancerous cells. This simple approach is useful in conveying a general overview of carcinogenesis, although the molecular and biochemical processes at each stage have been more thoroughly elucidated over the past decade or more. The first stage is known as initiation, which refers to a genetic mutation caused by some physical (for example, radiation) or chemical agent, usually from the environment. The mutation may remain dormant in the affected cell, possibly for years, until some signal or condition stimulates the abnormal cell to divide and multiply, eventually forming a mass of cancerous cells that are out of control and sapping energy from the surrounding healthy tissue or the entire body. The latter stage is known as proliferation or promotion of cancerous cells. This model is simplistic but instructive. One of its shortcomings is that not all mutagens are carcinogens and vise versa.

Today it is recognized that there are specific genes that appear to be vulnerable to mutation and the resulting mutations can lead to aberrant function. Numerous genes have been identified as oncogenes (cancer-causing genes) or proto-oncogenes—genes with potential to be mutated and transformed into oncogenes. Once the proto-oncogenes have been mutated or transformed, this can result in expression of other genes that promote cell proliferation (division) or inhibit apoptosis (programmed cell death of precancerous cells). In addition, there are tumor suppressor genes, such as p53, which bind to DNA and promote repair of mutated DNA or, if the damage is extensive, trigger signals for apoptosis. It is beyond the scope of this chapter to discuss all the genes, proteins, and processes involved in the tight regulation of cell division, which may result in cell proliferation or apoptosis. However, it will be useful to give a general overview and discuss the role of lipids in these processes.

There are many highly reactive chemicals produced during lipid peroxidation that may produce genetic mutations. The chances of this happening are quite small because most cells have a wide range of protective mechanisms to prevent any chemical attack on DNA, and to repair damage to DNA if an attack does occur (see figure 10.1). These systems are not perfect, and mishaps can occur—more so in some individuals than in others.

A mutation may take place in a region of DNA that has no apparent function and ultimately will have no bearing on the future of the cell. However, if the mutation takes place in a critical gene, there can be dire consequences. One possible outcome can be that the genetic mutation results in loss of certain control processes in the cell. This could lead to the cell ignoring or not recognizing control signals coming from other parts of the organism or tissue. Another possibility is that the cell can receive a signal for activation, but the genetic mutation may result in a defective or missing enzyme (or other protein) that would, if present, return the cell to its normal idle status. This results in the cell remaining permanently activated, even though activation signals are no longer being received. The state of permanent stimulation leads to cell division, and the mutation is passed on to subsequent generations of cells that acquire the altered gene for perpetual growth. This cell line proliferates on its own and eventually becomes a tumor.

There is a possibility of having a genetic predisposition to developing certain types of cancer as a result of having a specific gene that makes one more prone to cancer. Among the more widely studied examples of such oncogenes or proto-oncogenes are the *BRCA1* and *BRCA2* genes for breast cancer. Although these have been extensively studied in terms of heredity, because genetic markers are readily available to identify them, the function of the gene products of these oncogenes is not well understood.

The *BRCA1* gene product seems to be a protein associated with DNA repair (Hartl and Jones 1991). If an individual has an impaired system for DNA repair, that person will naturally be at greater risk of having permanent mutations in DNA and therefore at greater risk of developing cancer. It is not clear why this particular gene and its defective DNA repair system affects primarily breast tissue and not other organs. Other genes have been associated with other types of cancers, and in some ethnic groups the *BRCA2* gene may be associated with risk of prostate cancer in men but is not as important as other genetic markers (Melamed, Einhorn, and Ittmann 1997).

Hereditary traits such as oncogenes are passed from generation to generation through the germ cell line and make those offspring who inherit the trait more susceptible to certain cancers. However, malignancy requires a series of events to proceed from what is known as a preneoplastic (precancerous) stage to the malignant or cancerous stage. Preneoplastic cells must avoid apoptosis,

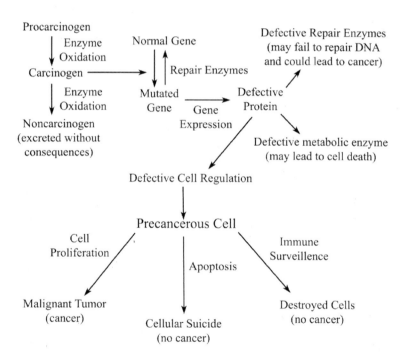

FIGURE 10.1. Carcinogenesis. Procarcinogens can be converted by oxidative metabolism (generally in liver) to carcinogens, which can enter the circulation and be taken up by cells in specific tissues to cause mutations in the DNA of those cells. The mutated gene may produce a defective protein that can have various consequences in the cell, either making the cell nonviable (it will not survive) or transforming the cell into a precarcinogenic cell, which may undergo apoptosis, be destroyed by the immune system, or survive and proliferate to form a cancerous tumor.

circumvent the need for certain growth factor signals from other cells, ignore regulatory signals from other cells, and avoid immune surveillance in order to survive and proliferate (Adams and Cory 1991). The proliferating cluster of cells also has to commandeer its own blood supply for nutrients (angiogenesis, see below) (Adams and Cory 1991). Some cells in a developing tumor may migrate to distant sites in the body via the blood, resulting in metastasis (growth of cancerous cells throughout the body). Most of these events depend on defective genes of various types.

There are many oncogenes, and their protein products give rise to a greater predisposition to specific types of cancer. Heredity is a major factor in determining one's risk for getting cancer, but to date there is little that can be done about the genes inherited at conception. It is possible to minimize or reduce one's risk for getting cancer by avoiding known environmental carcinogens, or at least taking some precautions to reduce exposure to these factors if such exposures are inevitable. If a person knows that he or she may have an inherited predisposition to cancer (for example, colon or breast cancer running in a family), then avoiding environmental carcinogens becomes more important, and greater attention should be paid to minimizing exposures. Unfortunately, perceptions of invulnerability in youth often run counter to good judgment with respect to awareness of environmental carcinogens, and many cancers have latency periods (lag times) of many years before cancer develops from the initial exposure.

Dietary Influences

Numerous studies have focused on dietary factors that may reduce or increase one's risk for getting cancer. It is generally recognized that naturally occurring antioxidants and a wide range of phytochemicals (chemicals produced by plants) can offer some protection against cancer (Wattenberg 1983). However, several clinical trials have found that antioxidant supplements are not always beneficial in protecting against cancer. In one study, vitamin E supplementation in cigarette smokers showed no effect, while beta-carotene (a plant-derived precursor for vitamin A) actually exacerbated the incidence of lung cancers in smokers by a small but significant amount (Albanes et al. 1996). This is not to say that beta-carotene promotes cancer, in general, but this study is mentioned to illustrate the complexities and inconsistencies of careful scientific studies.

It was mentioned earlier (chapter 5) that attempts to decrease risk of heart disease in subjects by feeding them more polyunsaturated vegetable oils to lower their blood cholesterol levels resulted in a significantly higher rate of cancers (Pearce and Dayton 1971). Studies done more than a half century ago showed high-fat diets increase the incidence of spontaneous and carcinogen-induced cancers in laboratory animals (Tannenbaum 1942). Many studies of diet and cancer in laboratory animals indicated dietary polyunsaturated oils, per se, would not cause cancers. However, if cancers were initiated with a known carcinogen and the animals were fed high polyunsaturated fat diets, there were often (but not always) more cancers than if the animals were fed predominantly saturated fatty acids.

In general, higher levels of fat in the diet increase the number of tumors in animals treated with a carcinogen, suggesting the dietary fats increase proliferation of cancerous cells. Polyunsaturated vegetable oils seem to promote cancer or stimulate proliferation of cancerous cells more than saturated fats, although changes in the profile of fatty acids in the diet (saturated versus polyunsaturated) do not always produce consistent results, leading to much confusion about what types of fats may increase risk of developing cancer.

In terms of dietary fats and cancer in humans, it is often difficult to make direct comparisons between or among studies because of study designs. Social, behavioral, and eating patterns of the study groups and biases of researchers involved may influence factors or omit important data that are relevant to observed outcomes. For example, a recent paper stressed in its conclusion that colorectal cancer is positively associated with high consumption of red and processed meat, yet a close inspection of the data shows that red meat alone was not associated with these cancers but processed meat was (Norat et al. 2005). The researchers combined red meat with processed meat and found there was still a positive association. However, anyone reading the article to get the general conclusions of the authors would come away with the impression that red meat of any kind would promote colorectal cancers. There seemed to be a clear bias of the authors when they drew their conclusions. They did not distinguish between processed and fresh red meats, even though they showed the data for such a distinction.

The U.S. Department of Agriculture pressured meat processors to reduce the amount of nitrites added as preservatives to bacon, ham, frankfurters, and other processed meats in the 1980s because nitrites are known to undergo reactions with amines to form nitrosamines under acidic conditions (such as in the stomach). Nitrosamines are known to be carcinogens. It is believed that nitrites in preserved meats are responsible for the high incidence of stomach cancer in countries that have a high consumption of preserved meats. The meat-processing industry claims there is no evidence that nitrites are associated with increased incidence of stomach or intestinal cancers. There are many environmental factors that may be influencing risk of cancers in the gastrointestinal tract, and cured meat is only one of them. Consumption of cured meats has also been associated with greater incidence of chronic obstructive pulmonary disease or asthma (Jiang et al. 2007).

When a study demonstrates the promotion of cancer by dietary fats in general, with no distinctions made between saturated fats and polyunsaturated oils, the effect may be due to the fats providing high levels of energy for cancerous cells to proliferate. Cancer cells seem to be quite efficient at extracting energy sources from the circulation and sapping energy from the rest of the body. It should come as no surprise when studies of the effects of dietary fats on cancer do not always produce consistent results, especially with the many different types of cancers occuring in genetically and culturally distinct populations where the investigations are done.

There may be some threshold above which dietary fats show no discrimination. When there is a high level of fat consumption, there may be no difference between saturated, monounsaturated, or polyunsaturated fats in terms of promoting cancers. But if total fat consumption is low, differences in the degree of saturation or unsaturation, as well as differences between omega-3 and omega-6 polyunsaturated fatty acids, may become more important. This has been emphasized by Japanese workers who pointed out that failure to observe any link between consumption of omega-3 fish oils and colorectal cancer in the American Nurses' Health Study, in contrast to favorable findings in a Japanese study, is due to the large amount of omega-6 fatty acids in the American diet (Tokudome et al. 2006). The ratio of omega-3 to omega-6 fatty acids in the American diet was between 1:100 and 1:25, whereas this ratio was about 1:20 to 1:10 in the Japanese Dietitians' Health Study. In other words, because the American population is eating far more omega-6 vegetable oils than the Japanese, differences in the amount of omega-3 fatty acids consumed by Americans are not going to affect the incidence of cancers very much.

The effects of altering the profile of dietary polyunsaturated fatty acids (PUFAs) turned out to be interesting in several studies. A few examples demonstrating differences in actions of dietary PUFAs from various sources illustrate the importance of eicosanoids formed from omega-3 versus omega-6 essential fatty acids in another set of diseases, cancers.

Dietary vegetable oils (rich in omega-6 PUFAs) promoted chemically induced mammary tumors in rats much more than diets containing predominantly saturated fatty acids (from coconut oil) or omega-3 PUFAs from fish oil (Braden and Carroll 1986). Dietary fish oils suppress other kinds of tumor cell growth in laboratory cell cultures relative to dietary vegetable oils (Karmali 1987). Bioactive eicosanoids derived from dietary essential fatty acids thus appear to be important with regard to tumor promotion and cell proliferation (McGiff 1987).

Cyclooxygenase-2 (COX-2) expression has generally been found to be increased in cancerous tumors relative to normal tissue. As mentioned earlier (chapter 9), the COX-2 form of this enzyme is inducible. Many studies have shown that frequent use of nonsteroidal anti-inflammatory drugs (NSAIDs) will lower risk of cancers. It is interesting to note that an active constituent of green tea, epigallocatechin-3-gallate, suppresses the induction of COX-2 without affecting the expression of COX-1 (Hussain et al. 2005). This may explain the therapeutic effects of green tea regarding cancers, as well as other afflictions associated with elevated levels of COX-2 (see chapter 9).

The role of eicosanoids in immune response (the immune surveillance system) must also be considered in the body's ability to check the growth of tumors (Janniger and Racis 1987). There are several cytokine (protein) signals that activate the immune system, but a host of lipid (eicosanoids and other lipids) signals are also involved. There is still much to learn about their interactions that result in successful or failed immune surveillance.

Angiogenesis and Apoptosis

In order for a tumor to grow, it must have an adequate blood supply to provide nutrients and oxygen for the rapidly growing cells. This requires formation of new blood vessels in the tumor, a process known as angiogenesis. New blood vessels are built from existing microvasculature (blood capillaries). Angiogenesis is important in several diseases in addition to tumor growth, such as rheumatoid arthritis and retinopathy (disease of the retina in the eye) arising from diabetes. New growth of blood vessels is controlled by numerous signals, including a cytokine known as vascular endothelial growth factor (VEGF). It turns out that prostaglandins produced by the inducible COX-2 enzyme found in the vascular cells can stimulate the expression of VEGF (figure 10.2). Hypoxia (low oxygen in the blood) and rapid consumption of oxygen and nutrients in tumor cells also stimulate the induction of VEGF. Several studies have shown that inhibition of COX-2 resulted in decreased tumor angiogenesis and diminished tumor growth in colorectal, prostate, and breast tumors, and decreased formation of secondary bone tumors resulting from the spread of small-cell lung carcinoma (Klenke et al. 2006).

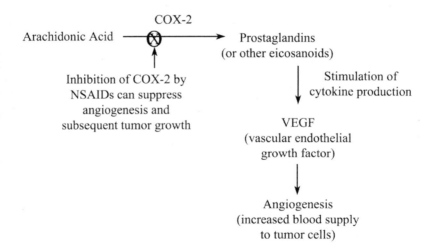

FIGURE 10.2. Angiogenesis can be inhibited by NSAIDs and COX-2 inhibitors through suppression of prostaglandin production from arachidonic acid.

Nonselective NSAIDs (aspirin and ibuprofen) gave similar reduction in breast cancer, relative to selective COX-2 inhibitors, if normal doses were taken with similar frequency (at least three times per week). On the other hand, low-dose aspirin or a regular dose taken less than twice per week was ineffective (Harris, Beebe-Donk, and Alshafie 2006). This study did not show any data for the selective COX-2 inhibitors at less than daily doses, but one would expect results similar to those for the nonselective NSAIDs. The failure of daily low-dose aspirin to suppress tumor growth is probably due to the fact that this small amount of aspirin is consumed by blood components (for example platelets) first, and little would find its way to the tumor sites.

Eicosanoids may also influence apoptosis of tumor cells. Levels of PGE_2 are very high in semen (chapter 3). It was thought that the prostate gland was producing PGE_2, hence the name prostaglandin, but it was later found that it is produced in the seminal vesicles. PGE_2 has been shown to stimulate cell proliferation and to increase levels of a protein known as Bcl-2, which suppresses apoptosis. Consequently, since NSAIDs inhibit PGE_2 production, they may suppress cell proliferation and increase apoptosis of aberrant cells (figure 10.3). Signaling factors and events involved in apoptosis are complex, but it suffices to say that inhibition of PGE_2 production by COX inhibitors (NSAIDs as well as selective COX-2 inhibitors) would allow more Bcl-2 to be produced, permitting apoptosis (death of tumor cells) to proceed (Sheng et al. 1998). There will be more on this below with regard to the effects of conjugated linoleic acid.

Organic peroxides, such as benzoyl peroxide (used as a free-radical-generating compound in the manufacture of plastics) and lauroyl peroxide (another lipid peroxide that was tested), are effective tumor-promoting agents (Slaga et al. 1981). Peroxides promote cancerous skin tumors by generating free radicals

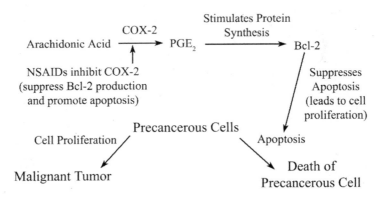

FIGURE 10.3. NSAIDs can promote apoptosis by suppression of prostaglandin E2 production and subsequent suppression of Bcl-2 production. Bcl-2 normally suppresses apoptosis, so less Bcl-2 would increase apoptosis.

that cause changes in cell membranes, which may affect signals coming from other cells. It would be logical to assume that fatty acid peroxides formed during lipid peroxidation would promote tumor growth in a similar way, but studies apparently have not been documented to show this.

Conjugated Linoleic Acid

The term *conjugated linoleic acid* (CLA) refers to a few isomers of normal linoleic acid, the omega-6 fatty acid that is most abundant in vegetable oils. These conjugated isomers generally have one trans double bond adjacent to a cis double bond. Normal or unconjugated linoleic acid has two cis double bonds separated by one saturated carbon or -CH_2- group. The structures of two common conjugated linoleic acid isomers are shown in figure 10.4, along with unconjugated linoleic acid. Although the trans double bond in CLA is analogous to those in trans fats found in partially hydrogenated vegetable oils, the position of a trans double bond next to a cis double bond in CLA imparts very different chemical characteristics as well as physiological effects. Such a juxtaposition of double bonds with a single bond connecting them is known as conjugation in chemical parlance.

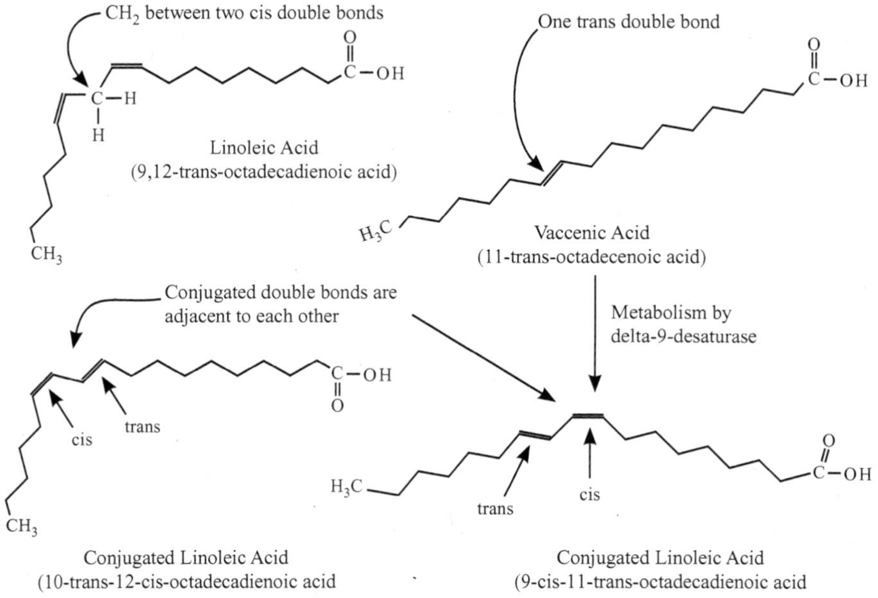

FIGURE 10.4. Structures of vaccenic acid, linoleic acid, and two of its conjugated isomers, 10-trans-12-cis-octadecadienoic acid and 9-cis-11-trans-octadecadienoic acid. Note the relative positions of the double bonds and cis versus trans orientation around the double bonds.

One particular trans fatty acid, vaccenic acid, is metabolized by the delta-9-desaturase enzyme (see chapter 2) in mammals and humans (figure 10.4) (Lock et al. 2004). The food industry claims there are trans fats in beef and dairy products like the trans fats found in partially hydrogenated vegetable oil, margarine, and shortening. However, not all trans fats are created equal. The trans fats resulting from hydrogenation of vegetable oils have an abundance of trans double bonds in positions 9 or 10 of the fatty acid chain, which prevents them from being metabolized by the delta-9-desaturase enzyme. Such trans fatty acids in margarine and shortening do not form conjugated acids such as CLA.

The trans fats in beef and dairy products arise from bacterial metabolism in the digestive system of cattle, goats, and sheep (Tsiplakou and Zervas 2008). The major trans fat found in these ruminants is vaccenic acid, which has the trans double bond in position 11 of the eighteen-carbon fatty acid (see figure 10.4). This allows the delta-9-desaturase enzyme in liver to add a cis double bond adjacent to the trans double bond that is already there. Inhibition of this desaturation metabolism in rats reversed any beneficial effects of vaccenic acid with respect to anticarcinogenic activity.

The anticarcinogenic properties of conjugated linoleic acid (CLA) were first revealed in Michael Pariza's laboratory at the University of Wisconsin, Madison, in the late 1980s. It was originally thought that the conjugated derivative was formed at high temperatures when beef, for example, was fried or grilled. It was later found that the conjugated isomer was present in uncooked beef and originates from bacterial sources in ruminant animals (Chin et al. 1992). CLA was eventually shown to be not only anticarcinogenic, but also to suppress atherosclerosis, diabetes, and formation of adipose tissue (it may suppress obesity). Early studies showed that CLA would inhibit tumorigenesis in the skin (Ha, Grimm, and Pariza 1987), mammary glands (Ip et al. 1991), and stomach (Ha, Storkson, and Pariza 1990) of experimental animals treated with specific carcinogens.

CLA inhibits carcinogenesis at each of the stages discussed earlier (Belury 2002). It appears to inhibit initiation in several ways (figure 10.5). The conjugated double bonds in CLA make it much less susceptible than the unconjugated double bonds of regular linoleic acid to lipid peroxidation. CLA inhibits free-radical oxidation reactions and metabolism of procarcinogens to their active carcinogenic form. It may also inhibit the interaction of carcinogens with DNA, thereby preventing mutations. CLA alters the profile of polyunsaturated fatty acids in cell membranes and the resulting production of eicosanoids from arachidonic acid, which would in turn affect the regulation of cellular activities in a variety of ways (figure 10.5, right side). CLA reduces proliferation of several cell types grown in culture and promotes apoptosis by decreasing the production of Bcl-2, the suppressor of apoptosis mentioned above.

As more is learned about this unusual unsaturated fatty acid found naturally in beef, goat, mutton, and dairy products, it will be interesting to see whether

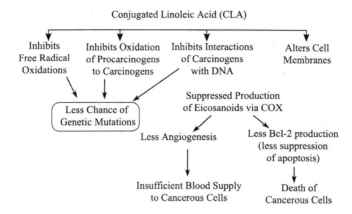

FIGURE 10.5. Anticancer effects of conjugated linoleic acid.

proliferation of its use in dietary supplements and well-controlled clinical trials will confirm the effects seen in laboratory animals regarding atherosclerosis and cancer. Controlled studies of CLA supplements in women regarding appetite suppression (Medina et al. 2000) and energy expenditure (Zambell et al. 2000) did not support claims of its effects on lowering body fat in mice and improvements seen in genetically obese rats. Perhaps it suppresses proliferation of fat cells to prevent obesity but is not effective in reversing the process; thus it may not remove fat cells once they are formed.

The strict dietary limitations that can be imposed in laboratory animal studies are quite different from the dietary choices of free-living humans who elect to take supplements, so it may be necessary to give human subjects more guidance with regard to dietary modifications when taking supplements for a particular health outcome. It may be necessary to reduce consumption of linoleic acid (omega-6 vegetable oils) and/or partially hydrogenated oils containing trans fats in order for beneficial effects of CLA to be noticed. In addition, high consumption of fructose from sweeteners may diminish any beneficial effects of CLA on obesity.

Studies of Dietary Fats and Cancer in Humans

The effect of dietary fats on tumor promotion is, in general, dose dependent. It is commonly thought that higher levels of fats result in greater tumor promotion. However, most dietary studies in humans rely on reporting the frequency with which subjects consume certain foods (a food-frequency questionnaire) and often neglect other dietary and behavioral factors in comparing dietary fat consumption (although tobacco smoking is often taken into account). For example, would a person who consumes a lot of salad, and consequently salad dressing containing polyunsaturated vegetable oil, and perhaps less butter, margarine, and meat (as a result of heeding dietary recommendations) have an overall healthier prognosis? Would that person have different behavioral patterns (such as exercise) compared with an individual who is less conscious about health, eating whatever is convenient at the time, not concerned about environmental exposures, and perhaps living a more sedentary life in terms of exercise? How are these differences accounted for in food-frequency question-naires?

Even in some animal studies, in which diet and other factors are carefully controlled over the duration of the study, lower incidences of tumors have been found with higher-fat diets (Lawrence et al. 1984; Nauss, Locniskar, and Newberne 1983). Since it is generally believed that high-fat diets promote cancer, such results do not fit well into the "low-fat is healthier" paradigm. Conflicting experimental results are frequently ignored or, if submitted, often not accepted for peer-reviewed publication, and the reasons for the unexpected results are seldom explored owing to lack of funding to pursue these avenues of research. Clearly the relationships of diet to cancer and other diseases are complex and frequently inconsistent. There are also numerous risk factors for various types of cancer that must be taken into account, such as, in the case of breast cancer, age, duration of reproductive life, history of estrogen replacement therapy, genetic predisposition, and body mass index, to name a few (Erickson 1998). Different risk factors may apply for other types of cancer, but some factors may generally apply to many types of cancer.

If there is some threshold above which high-fat diets promote cancer, and if most humans in a study population are consuming fats at levels well above that threshold, then any changes in the profile of fatty acids consumed will make little difference. For example, if the threshold were 25 percent of calories consumed as fat compared with 35 or 40 percent consumed as fat in the diet of average Americans, one would not expect to see much effect of subtle changes in fatty acid composition of the diet. Furthermore, Americans and Western Europeans typically consume levels of polyunsaturated fatty acids (6 to 10 percent or more of total calories) that are well above what is considered to be necessary to avoid an essential fatty acid (EFA) deficiency (less than 2 percent of calories). With these qualifying remarks made, it may be instructive to look at some studies.

In a European case-control study of breast cancer involving women from five countries, it was found that body mass index, family history of breast cancer, and older age at first childbirth were the strongest risk factors for development of breast cancer (Kohlmeier et al. 1997). Socioeconomic status, alcohol consumption, and postmenopausal supplemental hormone use showed no association with breast cancer. This study measured the profile of fatty acids in adipose tissue (from the buttock), which is generally considered a more reliable assessment of dietary fat consumption than food-frequency questionnaires, especially with respect to polyunsaturated and trans-fatty acids. It was found that trans-fatty acids (mostly trans 6 and trans 9 isomers of oleic acid) in adipose tissue were associated with breast cancer (odds ratio of 1.4 or 40 percent higher risk for high levels versus low levels of trans-fatty acids in fat tissue). What is more interesting is that trans-fatty acids had the greatest influence on breast cancer development in the group with the lowest levels of polyunsaturated fatty acids (PUFAs) in adipose tissue. There was no association between trans-fatty acids and cancer in the group with the highest levels of PUFAs in adipose tissue. It would seem that high levels of PUFAs in the diet give rise to more cancers and that additional trans-fatty acids on top of that are not augmenting tumor promotion very much. However, when PUFA consumption is low, trans-fatty acids may show a significant effect because there is much less influence from PUFAs. This study controlled for a wide range of risk factors. Other studies have produced inconsistent results, but in many cases other risk factors were not taken into account.

Although there seems to be a general perception among most scientists that PUFAs from vegetable oils promote cancers, one review and meta-analysis (Zock and Katan 1998) of several studies indicates that dietary linoleic acid (the major omega-6 fatty acid found in vegetable oils) is not a risk factor for human cancers (breast, colorectal, and prostate). Some studies show a significant decrease in risk of cancer with high consumption of polyunsaturated oils, while others may show a significant increase. Most of the studies show no significant relationship between linoleic acid (or polyunsaturated vegetable oil) consumption and cancer in these three organs. In most studies, the majority of subjects may have been well above the threshold for PUFA consumption, so any additional intake of PUFAs may have had no effect on cancer progression. There is good support for this notion from comparison of studies done in Japan versus the United States (see the section above on dietary effects). The omega-6 PUFAs seem to affect predominantly the promotional stage of cancer, which would mean a more rapid progression of the disease once initiation has taken place.

Several studies have shown that ethnic groups that consume more fish have a lower incidence of cancers (Larsson et al. 2004). Japanese women who decreased their consumption of fish and increased their intake of vegetable oils over the past decades (shifting toward more Western diets) have increased their rates of breast cancer (Lands et al. 1990). However, case-control and prospective cohort studies within various ethnic populations or countries often produce less convincing results, with some showing a favorable effect of omega-3 fatty acid consumption but many showing no effect.

Again, the lack of effect in a majority of studies may be due to the relative ratio of omega-3 to omega-6 fatty acids in the diet. Differential effects of omega-3 versus omega-6 bioactive eicosanoids have been discussed throughout this book. The bioactive eicosanoids, as well as fatty acids, are involved in signals for expression of a wide range of cytokines that are involved in cell proliferation, angiogenesis, apoptosis, metastasis, and immune surveillance, all of which influence the ultimate outcome of the body's battle with cancer.

Statin Drugs and Cancer

Lipid-lowering drugs (statins) are prescribed to lower blood cholesterol levels, especially in people at high risk for developing heart disease. It is often said that low blood cholesterol levels increase one's risk of developing cancer. This statement can be as misleading as saying that anything that raises blood cholesterol will increase one's risk of heart disease. Statins are very effective at lowering blood cholesterol, so one might expect that they could increase risk of cancer, but this does not appear to be the case. Feeding animals extremely high levels of statins (as with many substances) can increase the frequency of cancers in those animals. However, several studies in animals in which moderate doses were used show that statins can have an anticarcinogenic effect (Jakobisiak and Golab 2003).

Lovastatin has been shown to suppress proliferation of several types of tumor cells in culture, including breast, prostate, lung, pancreatic, and gastric carcinomas, to name a few. Lovastatin can also inhibit growth of normal cells, but this effect is less pronounced owing to slower proliferation of normal cells compared with tumor cell lines.

Lovastatin, as well as other statins, can induce apoptosis in many types of tumor cells. The mechanism for inducing apoptosis is through inhibitory effects on the enzyme responsible for cholesterol synthesis, HMG-CoA reductase, but not because of its effects on cholesterol levels. The product of the reaction catalyzed by HMG-CoA reductase is mevalonate (or mevalonic acid), which is used to make not only cholesterol but several other lipid products, such as farnesyl pyrophosphate, an isoprenoid (see figure 10.6). Isoprenoid compounds, such as the farnesyl group, are involved in regulating cell growth and division.

Growth factors stimulate growth in cells by binding to specific receptors on the cell membrane. This signal is transmitted in the cell via a family of proteins known as Ras (there are many different Ras proteins). Prenylation (attaching isoprenoids such as farnesyl) to these proteins is an essential step in transformation of a normal cell into a tumor cell. Prenylation does occur in normal cells, but the isoprenoids remain attached to the proteins only transiently and are removed to restore the cell to its quiescent state. When HMG-CoA reductase is inhibited by statins, the diminished production of mevalonate results in less farnesyl pyrophosphate, hence less conversion of the Ras proteins to their prenylated forms and less cell proliferation.

The Ras family of proteins are produced from proto-oncogenes, which are highly susceptible to mutations that make them oncogenes. The oncogenes

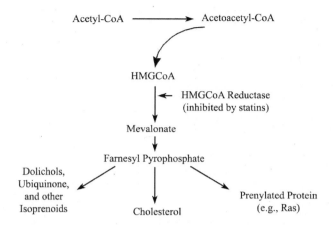

FIGURE 10.6. Production of mevalonate is inhibited by statin drugs, leading to diminished production of cholesterol, prenylated proteins, and other isoprenoid products.

produce Ras proteins that become permanently activated when a growth factor signal is received. Normally, the Ras proteins become activated for a short time and then are inactivated, but the genetic mutations result in expression of proteins that remain permanently activated, leading to rapid cell proliferation. More detailed descriptions of these processes can be found in textbooks describing the biochemistry and/or molecular biology of cancer.

Lovastatin also induces apoptosis in a wide range of tumor cells (Erickson 1998). Administration of mevalonate, the immediate product of HMG-CoA reductase, which is inhibited by statins, resulted in suppression of statin-induced cell death and promotion of cell proliferation. Lovastatin was also shown to inhibit the expression of the Bcl-2 protein that suppresses apoptosis (see above). Lovastatin has also been shown to inhibit several essential steps in metastasis (Erickson 1998). These antitumor actions of statins are independent of their effects on cholesterol synthesis and blood cholesterol levels.

Conclusions

Although there is evidence that omega-3 fatty acids found in fish oils can have beneficial effects with regard to particular types of cancer, several studies that attempted to show beneficial effects by using dietary supplements have been less compelling. A better understanding of the role of various lipids, cytokines, and other signaling agents in carcinogenesis allows one to make more enlightened decisions regarding what effects dietary fats may have on cancer. However, there are still many uncertainties regarding the impact of polyunsaturated vegetable oils on carcinogenesis.

The lack of any consistent effect with respect to levels of dietary linoleic acid could be due to the fact that most Western and Northern Hemisphere populations consume this omega-6 fatty acid well above the levels needed to satisfy the nutritional requirements for this essential fatty acid. Consumption of fats beyond a certain threshold, whether of polyunsaturated fatty acids or total fats, may result in little or no effect when changing the overall profile of fatty acids consumed in dietary fats and/or taken as supplements. In addition to some clear evidence for beneficial effects of omega-3 fish oils, there is also evidence that COX inhibitors can suppress cancers. These two facts, as well as many in vitro studies, indicate that prostaglandins (and perhaps other eicosanoids) are playing a role in the promotion of tumors.

It was perhaps surprising that conjugated linoleic acid (CLA), which is found in beef, mutton, and dairy products, has anticarcinogenic properties. Although conjugated linoleic acids have a trans double bond, they appear to be handled quite differently from the trans-fatty acids found in partially hydrogenated vegetable oils. The presence in body fat of trans-fatty acids from hydrogenated vegetable oils was shown to be associated with breast cancer in a European study. The association was strongest when there were lower levels of polyunsaturated fatty acids in adipose tissue.

Statins may suppress cancers by inhibiting the enzyme HMG-CoA reductase, an effect that is not due to decreased cholesterol. Inhibition of this enzyme results in decreased production of several lipids in cells that influence proliferation of cancerous cells.

Clearly, dietary fats have effects on cancer, but there is still much more to be learned about how they elicit these effects. Scientists and health professionals have many biases when it comes to dietary recommendations regarding cancer, as they do with many other diseases. This chapter has focused primarily on lipids, especially polyunsaturated fatty acids, in carcinogenesis. But as with most health issues, genetics plays an important role.

11

Neurological Development, Memory, and Learning

The human brain is, without a doubt, the most complex structure we can contemplate. It seems ironic that we use this same organ in its utmost capacity to begin to understand its own complexity. Unlike most organs of the body that are relatively homogeneous in their structure, composition, and functions, the brain is extremely heterogeneous. It consists of billions of cells in numerous, discrete regions that have been mapped through medical history with assignments for sensory, motor, and cognitive functions.

Biomedical research has elucidated mechanisms by which neurons send and receive messages to elicit the numerous processes of the brain. The brain integrates bodily functions—a characteristic of nervous systems in the simplest organisms. It is also responsible for memory, learning, and more advanced thought processes that are often considered unique to humans. It may be surprising to find that much of our understanding about learning and memory has come from studies in invertebrates; the California sea slug, *Aplysia*, has been used as a simple model system for probing the molecular mechanisms for memory (Lee et al. 2008).

Overview of Brain Structure and Function

The brain is composed of two major types of cells: neurons and glia. Neurons have held the spotlight in neuroscience for the past century because they are considered to be responsible for all the active communicative processes that are characteristic of the brain. Glia play a supportive role and are generally placed into three major categories known as astrocytes, oligodendrocytes, and microglia. Microglia have characteristics of immune cells (monocytes and macrophages) and are responsible for cleaning up damaged brain tissue. Oligodendrocytes form the myelin sheath surrounding the axons of many neurons in the brain to effect more rapid communications. Astrocytes play a wide range of roles. They segregate neurons from one another, remove excess neurotrans-mitters and waste products from interstitial spaces (between cells), modulate synaptic activity and growth, and supply neurons with nutrients for their meta-bolic processes. Studies in the past decade or so suggest that glial cells play a more active role than they were given credit for earlier (Miller 2003).

Neurons account for only about one-fourth of brain weight (Bourre 2004); there are more glia than neurons in the brain, but there are also fluid-filled ventricles. Neurons come in a wide variety of sizes, shapes, and functions. They have dendrites and cell bodies, which receive signals from other neurons, and they send signals via their axons (see figure E.3 in appendix E). There may be thousands of dendrites and synaptic connections for incoming signals on each neuron, but there is a single axon extending from the cell body. The axon may be branched to form hundreds or even thousands of synapses with many other neurons. When a neuron fires (sends its signal by releasing its neurotransmitter from the axon terminals), all axon terminals release transmitter at the same time as a result of axon depolarization. However, there may be local signals at specific terminals that can suppress transmitter release. Such localized effects give rise to fine-tuning of neuronal signals.

Neurons are often categorized according to the type of neurotransmitter they release for communication with neighboring neurons. There are several classic monoamine neurotransmitters, such as acetylcholine, dopamine, norepinephrine, and serotonin, as well as the amino acid neurotransmitters, glutamate and GABA (gamma-amino butyric acid). These few neurotransmitters make up the bulk of neurotransmitters in the brain. Consequently their proper-ties and mechanisms have been more thoroughly studied. In addition, there are numerous peptide neurotransmitters, which are far less abundant but can have very potent actions and play important roles. Peptide neurotransmitters are short chains of protein, ranging from as few as three amino acids per chain to forty or more. The low abundance of peptide transmitters, as well as their close physical and chemical relationship to proteins, make it much more difficult to study them compared with the amine and amino acid neurotransmitters.

In the mid-twentieth century scientists developed techniques to identify amine transmitters and found that individual neurons contained only one type of amine transmitter, either dopamine, norepinephrine, acetylcholine, or serotonin, but not two of them together. This led to the hypothesis that one neuron would have only one neurotransmitter. If a neuron stored and released dopamine, it was called a dopaminergic neuron. Similarly, there were cholinergic (for acetylcholine), noradrenergic (for noradrenaline, which is also known as norepinephrine), and serotonergic neurons. Later it was discovered that some neurons contained peptides along with the amine (or amino acid) transmitters, although dopaminergic neurons in some regions may contain one peptide transmitter, while dopaminergic neurons in another region of the brain might have a different peptide transmitter. Such combinations of amines with peptide transmitter seem to accommodate greater flexibility in terms of fine-tuning signals.

Finally, there are lipids that function as signal transmitters in the brain. Some lipid transmitters defy our traditional model of neurotransmission being in one direction. Traditionally the axon terminal of a neuron was thought to release transmitter that binds to a receptor across the synapse (usually on dendrites or the cell body of the receiving neuron). Instead, many lipids act as retrograde transmitters, meaning that they go from the postsynaptic (receiving) cell to the presynaptic (sending) cell across the synapse. In other words, they travel against the flow of traffic of amine or amino acid transmitters. Many of these lipid messengers are eicosanoids that have been discussed throughout this book, such as prostaglandins and endocannabinoids, as well as docosatrienes (derived from DHA) and platelet activating factor (PAF). These lipid messengers will be discussed in more detail with regard to specific brain functions or in relation to brain disorders.

There are a large number of neurotransmitters (amino acids, amines, peptides, and lipids) to track in this myriad of neurons and synaptic connections. Each transmitter may have several different types of receptors on receiving cells, which may cause different cells or even different parts of the same cell to respond differently to one neurotransmitter. For example, the excitatory neurotransmitter glutamate has at least three types of receptors linked to ion channels that cause depolarization of receiving cells. Glutamate also has several different receptors linked to G-proteins in the receiving cells (see chapter 4), which alter other proteins or enzymes, and hence metabolism.

Several different types of receptors have been characterized for all the major neurotransmitters, as well as for many peptide and lipid transmitters. So, if the number of different transmitters in the brain seems overwhelming, the number of possible combinations with receptors and subsequent responses becomes absolutely mind-boggling. A more complete overview of neuronal mechanisms is given in appendix E.

Essential Fatty Acids and Brain Development

There has been a large body of research over the years to determine what fatty acids are essential for the central nervous system and what roles they play in growth, development, structure, and signaling. The brain is a unique organ in terms of its composition, structure, and function. It has the highest percentage (by mass) of lipids of any organ, other than adipose tissue; nearly two-thirds of its dry mass is lipid. A major difference between brain and adipose lipids is that brain tissue is mostly phospholipids, whereas adipose tissue is mainly triglycerides. About 35 percent of lipids in brain tissue are long-chain (more than eighteen carbons) polyunsaturated fatty acids (LCPUFAs). The eighteen-carbon polyunsaturated fatty acids, linoleic acid and linolenic acid, are practically nonexistent in the brain.

DHA constitutes about 7 percent of total fatty acids in the human cerebrum and nearly 20 percent of fatty acids in the retina of the eye. The other main fatty acids are palmitic acid (a saturated fatty acid) and the monounsaturated fatty acids, oleic and palmitoleic acids. The saturated and monounsaturated fatty acids account for about 65 to 75 percent of fatty acids in the brain and retina (Yavin, Brand, and Green 2002). DHA constitutes about one-third of the LCPUFAs in the brain, with arachidonic acid accounting for another one-third. There is a relatively large amount of docosatetraenoic acid (22:4 omega-6) and relatively low levels of eicosapentaenoic acid (20:5 omega-3; EPA) in brains of humans that typically consume Western diets (Bourre 2004). The relative amounts of the latter two LCPUFAs are probably a reflection of low consumption of omega-3 fatty acids and relatively high consumption of omega-6 fatty acids in this population.

Lipids in the brain are not used for energy but are there mostly for structural purposes (in membranes) or for signaling after their metabolism to bioactive messengers. LCPUFAs are found almost exclusively in phospholipid membranes and as cholesteryl esters, with trace amounts as free fatty acids or metabolites of phospholipids that have lost their polar phosphate group.

Unanticipated Antioxidant Properties
of Long-Chain Polyunsaturated Fatty Acids

Brain tissue has the highest concentration of docosahexaenoic acid (DHA) in the body, except for the retina of the eye (which is part of the central nervous system). This may seem surprising in view of the greater expected susceptibility of DHA to lipid peroxidation—the more double bonds there are in a PUFA, the greater its chances of reacting in free-radical oxidation reactions. However, many studies have shown that increased consumption of DHA by lab animals usually does not increase levels of lipid peroxidation and, in fact, decreases lipid peroxidation in many organs, especially the brain (Yavin, Brand, and Green 2002).

There is an explanation for this contradictory effect of DHA in the brain as well as in certain other tissues such as the retina, heart, and testes, where DHA is also abundant. DHA is bound primarily to phosphatidylethanolamine and phosphatidylserine phospholipids. The phosphatidylethanolamine fraction contains a special class of phospholipids known as plasmalogens, which contain an ether linkage to a long-chain alkyl group in addition to the ester linkage of DHA (or other LCPUFAs). One of the more abundant of these ether-linked alkyl groups contains a double bond adjacent to the ether linkage, making this double bond both more reactive than a double bond in oleic acid and closer to the surface of the phospholipid membrane. The location and its chemical reactivity make this double bond more susceptible to attack by free radicals migrating to the cell membrane from the cytoplasm or other aqueous compartments where these radicals are usually generated. DHA dietary supplements may increase production of plasmalogens, thereby offering this unexpected antioxidant protection (Kuczynski and Reo 2006).

The white matter of the brain is enriched in plasmalogens, especially the myelin sheath formed by oligodendrocytes. The importance of plasmalogens is evident from the wide range of disorders in humans—such as Zellweger syndrome, infantile Refsum disease, rhizomelic chondrodysplasia punctata, Alzheimer's disease, and Down syndrome—that are associated with defective production of these phospholipids (Nagan and Zoeller 2001). Although the exact function of plasmalogens is unclear, they may play an important role in preventing oxidation of lipid membranes. Enrichment of LDL particles or other phospholipid mixtures with plasmalogens makes them more resistant to free-radical-induced lipid peroxidation (Hahnel et al. 1999). More traditional antioxidant systems tend to be less abundant in the brain, such as catalase, superoxide dismutase, glutathione peroxidase, and glutathione (Mizuno and Ohta 1986). It would make sense that a special lipid antioxidant system is operational in this vital, lipid-rich organ.

DHA Is Essential for Brain Development

There is efficient accretion of DHA in the brain during fetal and postnatal development, when myelin is being produced and synaptic connections are being formed. The level of DHA is highly conserved once it is there. If DHA is not available during this window of rapid brain growth (often called the brain growth spurt) as a result of maternal dietary deficiency in omega-3 fatty acids, the brain will substitute docosapentaenoic acid (DPA) derived from omega-6 fatty acids that are generally abundant in Western diets.

The rapid rate of brain growth in humans begins in the third trimester of pregnancy and continues for about two years after birth. Formation of neurons (neurogenesis) takes place mostly between the fourteenth and twenty-fifth weeks of gestation. The ensuing brain growth spurt begins with glial cell formation, which aids formation of synapses (synaptogenesis), and this is followed by myelination of neurons. Synaptogenesis continues throughout life where existing neurons form new synaptic connections in the process of learning and memory.

It is important to keep in mind that formation of new synapses is crucial for long-term memory and learning (see below). The newly formed synapses are composed of specialized areas of membranes on adjacent cells that are juxtaposed for efficient and rapid communication with one another. Each neuron may form hundreds or thousands of synaptic connections with different cells, although there may be multiple synapses between the same two cells. Formation of new synapses requires a supply of fatty acids to make the phospholipid membranes, and DHA appears to be critical in this context.

During the brain growth spurt, there is an accrual of DHA, due to both increases in overall brain mass and abundance of DHA relative to other fatty acids. The requirement for DHA in the fetus and newborn infant is substantial during this period, with an estimated accretion rate of 35 milligrams DHA per week from the beginning of the third trimester of gestation until the end of the first year of life—estimates range from 3.6 to as much as 76 mg per week (Lauritzen et al. 2001).

Although the fetus is capable of forming DHA from linolenic acid, much of its DHA appears to be formed by the mother and subsequently transferred to the fetus from the mother's blood, via the umbilical cord and placenta. When it is being transferred from the mother's blood, DHA is bound to phospholipids such as phosphatidyl choline, which delivers much-needed choline to the fetus as well. Linoleic and linolenic acids are not incorporated into the brain or retina to any significant degree. They are either elongated and desaturated to form LCPUFAs or, when in excess, metabolized to saturated and monounsaturated fatty acids (see below) that are much less susceptible to free-radical oxidation. This suggests that LCPUFAs play a unique role, more than simply increasing fluidity of cell membranes.

Maternal Control of Polyunsaturated Fatty Acids

The most abundant fatty acid in brain phospholipids is palmitic acid (C16:0), which is a saturated fatty acid and the major product of the fatty acid synthase complex in humans and mammals. The relative amount of saturated fatty acids in most tissues of the body generally ranges from 40 to 50 percent, with palmitic acid predominating (Bourre 2004). When labeled linoleic acid and linolenic acid were fed to pregnant rhesus monkeys, these PUFAs were metabolized primarily to palmitic acid prior to incorporation into phospholipids of the brain and retina of the fetus (Sheaff Greiner et al. 1996). On the other hand, labeled DHA was incorporated into the fetal brain and retina, with relatively little of it ending up in the saturated and monounsaturated fatty acid pool. This indicates that the brain and retina require predominantly palmitic acid, in addition to DHA and arachidonic acid, as the major fatty acids in the phospholipids of these neurological tissues. The ratio of linolenic acid to DHA in maternal blood was seventeen to one, whereas this ratio was about two to one in fetal blood. This indicates a preferential uptake of DHA from the maternal plasma by the fetus.

These are indeed most curious findings: they indicate that the brain and retina carefully control the composition of their phospholipid membranes. Perhaps most surprising is the fact that excess PUFAs are recycled to saturated and monounsaturated fatty acids in brain tissue. The latter are not susceptible to lipid peroxidation, which can wreak havoc in any tissue. The specific pathway for fatty acid recycling was not shown, but it would appear that eighteen-carbon PUFAs were metabolized by normal mitochondrial and peroxisomal beta-oxidation. Most of the acetyl-CoA produced in this pathway would be reused by fatty acid synthase to produce palmitic acid. Some of the acetyl-CoA produced in the mitochondria would undergo complete oxidation in the Krebs cycle to form the ATP needed for fatty acid synthesis. However, much of the malonyl-CoA used for fatty acid synthesis is derived from peroxisomal acetyl-CoA (Reszko et al. 2004).

Are Omega-3 Supplements Necessary for Infants and Children?

The source of DHA, or its omega-3 PUFA precursors, for the developing fetus and infant during the brain growth spurt is obviously from maternal body stores and diet during pregnancy and lactation. There is concern about whether contemporary Western diets contain sufficient levels of omega-3 fatty acids to provide children with ample amounts of this vital fatty acid for proper neurological development (Simopoulos 1991). DHA levels in maternal plasma decrease during the course of pregnancy. Furthermore, women who have had multiple pregnancies have lower plasma levels of DHA (relative to other fatty acids) than women in their first pregnancy (Al, van Houwelingen, and Hornstra 1997). This is despite the fact that a woman accumulates DHA in body stores throughout life. Highly efficient transfer of DHA across the placenta ensures that the fetus will receive maximum amounts of available DHA from the mother. However, if DHA availability is low, there could be less than optimal amounts incorporated into nerve tissue during development.

Some studies have shown that pregnancies result in better outcomes, in terms of duration of pregnancy and infant birth weight (less risk of premature birth), when the mother receives supplements of omega-3 fatty acids or consumes seafood regularly during later stages of pregnancy (Smuts et al. 2003; Olsen and Secher 2002; Olsen et al. 1993). Such results suggest that premature births result from inadequate maternal levels of DHA. In addition, DHA supplementation of infant formula can have a significant effect on neurological, visual, and intellectual development of premature infants, more so than for full-term infants (Heird and Lapillonne 2005). The greater effect of DHA supplementation on preterm infants can be explained by the fact that full-term infants have obtained substantially more DHA from the mother prior to birth, including significant reserves of this all-important fatty acid in adipose tissue and liver. DHA in fat stores of full-term infants can be utilized for postnatal brain development if there are less than adequate supplies in their diet during the first year.

Clearly DHA plays a vital role in neurological development of the fetus and infant, but there still appears to be some controversy over whether DHA supplements are necessary in commercial infant formulas (McCann and Ames 2005). One study in the early 1990s showed that preterm infants supplemented with fish oil had significantly less weight gain than infants receiving a standard unsupplemented formula during the first year of life (Carlson et al. 1993). It was found that their weight at twelve months (corrected age) correlated well with plasma levels of arachidonic acid measured at several times during the first year of life. A subsequent study using formula supplemented with low-EPA (but high-DHA) fish oil showed no significant differences in growth compared with unsupplemented formula (Carlson, Werkman, and Tolley 1996). The reasoning behind the latter study was that EPA in the earlier fish oil supplements was at a level high enough to inhibit metabolism of omega-6 fatty acids, particularly the conversion of linoleic acid to arachidonic acid.

Arachidonic acid is important in a wide range of effects in the body as a precursor for bioactive eicosanoids. Another study performed in the early 1990s did not show differences in growth in preterm infants receiving omega-3 supplemented versus unsupplemented formulas (Uauy et al. 1994). This was most likely because the supplements also contained soy oil, which would supply linoleic acid needed for arachidonic acid synthesis. A recent analysis of fourteen trials of LCPUFA supplementation of infant formulas concluded there was no evidence that LCPUFA supplements have any (positive or negative) effect on growth of full-term infants (Makrides et al. 2005). The positive effects of omega-3 supplements seem to be most prevalent for preterm infants. However, there would be no detrimental effect of omega-3 supplements for all infants, provided adequate levels of omega-6 fatty acids are included in their diet. In addition, growth or body weight should not be the only criterion used to assess efficacy of omega-3 formula supplements. A summary of the effects of dietary essential fatty acids is given in table 11.1.

Learning and Memory

One of the more fascinating advances in neuroscience is elucidation of the biochemical events associated with learning and memory. It is generally agreed that memory arises from development of new synaptic connections between existing neurons. The process is generally called neurite outgrowth or synaptic plasticity. Great strides have been made in isolating and characterizing numerous proteins and processes involved in memory, which is prerequisite to learning.

Dozens of proteins have been implicated in memory development. One event that lies at the heart of memory is long-term potentiation (LTP). LTP refers to cellular responses that are augmented or enhanced as a result of specific types of neuronal activity. LTP is generally driven by the excitatory neurotransmitter glutamate, although modulatory neurotransmitters also play a part (dopamine, acetylcholine, and serotonin). Numerous biochemical events are involved, along with associated proteins; the overall picture is still not well understood. For example, N-methyl-D-aspartate (NMDA) receptors for glutamate are known to play a role in LTP, but they can also induce an opposing effect known as

TABLE 11.1.
Important considerations regarding
essential fatty acid status of mother and child

Mother	Fetus/infant
Pregnancy	
Accumulates omega-3 fatty acids throughout life, with some reserves in adipose fat.	Has high demand for DHA during third trimester of pregnancy for neurological development.
Converts linolenic acid to DHA much faster than fetus.	Able to convert linolenic acid to DHA, but not as fast as mother can.
Multiple pregnancies can deplete stores of omega-3 fatty acids.	Preferential uptake of DHA across placenta from mother's blood to fetus.
Omega-3 fatty acid supplements during pregnancy can improve outcome—duration of pregnancy and birth weight (lower risk of premature birth).	Fetus will store excess DHA in adipose fat and liver in latter weeks of full term pregnancy; preterm infants will have less DHA stored.
Lactation	
Human breast milk has the highest DHA content of any land mammal.	Infant continues to get DHA from breast milk; formula should have adequate supply of omega-3/DHA.
Adequate dietary omega-3 intake during lactation will ensure adequate transfer to breast-feeding infant.	Formula milk should have adequate (not excessive) amounts of both omega-3 and omega-6 fatty acids.
Most omega-3 fatty acids are converted to DHA in mother's liver prior to transfer to fetus or milk.	High demand for DHA for neurological development continues for at least first year of life.

Source: Compiled by the author from information in the sources cited in the text.

long-term depression (LTD), which is characterized by diminished response to neuronal activity (Sheng and Kim 2002).

A key component of LTP is phosphorylation of proteins through activation of various protein kinase enzymes. Protein kinases are activated by various second messengers in cells including cyclic-adenosine monophosphate (cyclic AMP), calcium, calmodulin, and diacylglycerol, to name a few. The protein kinases use adenosine triphosphate (ATP) as the source of phosphate that is attached to proteins. The mechanisms involved in memory are only beginning to be worked out, and the role of lipids will be discussed here.

It seems LTP can account for short-term memory, whereby one can remember something for a few minutes but, without reinforcement, that memory is fleeting and disappears (such as a telephone number). This is likely due to removal of phosphate from existing proteins before any new proteins are synthesized by gene expression. Gene expression is required to make proteins that constitute new synaptic connections for longer-term memory, and phosphorylated proteins are involved in regulating that gene expression.

Postsynaptic receptors at glutamate synapses are attached to large protein complexes that are readily visible in the electron microscope and have come to be called the postsynaptic density or dendritic spine. This protein machinery is responsible for helping to regulate the strength or intensity of synaptic transmission. When rats were fed high-omega-3 fatty acid diets, compared with control diets, there were noticeable alterations in gene expression, with increased expression of some genes and suppression of others (Kitajka et al. 2002). It would be difficult to draw any conclusions from these data alone, but they show that dietary omega-3 fatty acids have an effect on a wide array of gene expression, including many that were not measured in this study.

The neurological basis for long-term memory and learning, as mentioned earlier, is formation of new synapses. This requires additional branching of axons and dendrites, much like branching of trees to produce more fruit. Neuronal cells grown in culture can be stimulated to produce more axonal branching. These cells will produce longer axon branches and more of them when there is an abundance of DHA in the growth medium (Calderon and Kim 2004). Other LCPUFAs did not stimulate neurite growth in these neurons, including arachidonic acid and docosapentaenoic acid (another omega-6 LCPUFA). The latter replaces DHA under conditions of extreme omega-3 fatty acid deficiency in growing animals, which suggests that omega-3 fatty acid deficiency may have an effect on axonal branching and, hence, memory and learning. It may be that DHA induces formation of antioxidant plasmalogens in these newly formed neurites, although this was not measured. Clearly DHA has an effect on synaptic plasticity, which would be expected to affect learning and memory. However, studies of the effects of DHA supplementation on cognition and behavior have not shown a convincing impact on normal individuals (see below).

Several neuropeptide transmitters have also been implicated in learning and memory, including vasopressin (antidiuretic hormone), oxytocin, angiotensin II, and insulin-like growth factors, to name a few. The large number of studies investigating the relationship of vasopressin to memory and learning stems from work in the late 1970s that showed vasopressin could improve memory. A remarkable case showed reversal of trauma-induced amnesia with vasopressin treatment (Oliveros et al. 1978).

Relatively high levels of leukotriene C_4 (LTC_4, a component of slow-reacting substance of anaphylaxis derived from arachidonic acid; see chapter 9) are found in the hypothalamus and median eminence of the brain. LTC_4 synthase, the key enzyme for LTC_4 production, is localized in the same regions as vasopressin neurons. This could explain a relationship between arachidonic acid (or omega-6 PUFAs) and memory, although the role of LTC_4 in memory or other cognitive functions is not clear. It is possible that omega-3 fatty acids get metabolized to the corresponding LTC_5, which may have distinct actions relative to LTC_4. Much more work is needed to give a clearer picture of interactions of these peptide and lipid signals, along with the more conventional neurotransmitters, in the overall processes of memory, learning, and behavior.

Sleep and Memory Consolidation

Sleep is a natural and necessary physiological state that rejuvenates the mind and body. Studies have shown that sleep deprivation impairs long-term memory and synaptic plasticity (students pulling all-nighters before exams should take notice). It was found that sleep deprivation causes increases in prostaglandin D_2 and 2-arachidonyl glycerol (2-AG, an endocannabinoid) and a decrease in prostaglandin E_2 (McDermott et al. 2003). It has also been shown that amide derivatives of certain fatty acids (for example, cis oleamide from oleic acid) can be isolated from cerebrospinal fluid of sleeping animals and will induce sleep when injected into awake animals (Cravatt et al. 1995).

Anandamide (an endocannabinoid formed from arachidonic acid) also induces sleep (Chen and Bazan 2005). The two predominant endocannabinoids, anandamide and 2-AG, fluctuate in opposing cycles with the diurnal rhythms of the body. Anandamide is low and 2-AG high during daylight, and vice versa during darkness. It was suggested that sleep deprivation and its associated elevation of 2-AG may be involved in the disruption of synaptic plasticity and long-term memory, although definitive proof of this relationship has not yet been reported. However, it is well known that exogenous cannabinoids (marijuana) can suppress memory.

Another lipid messenger, platelet activating factor (PAF), is involved in long-term potentiation (LTP) through its actions as a retrograde messenger that migrates back to the presynaptic neuron and augments release of glutamate at excitatory synapses (Chen and Bazan 2005). PAF is formed in the postsynaptic cell and migrates back to the presynaptic cell (as a retrograde lipid messenger) and increases the release of glutamate. The enhanced excitatory action of glutamate then triggers postsynaptic responses, such as activation of a protein kinase (CaMKII via calcium influx through NMDA receptors) and the cascade of events caused by this kinase enzyme attaching phosphate to a host of postsynaptic cellular proteins involved in learning and memory. This could promote the migration of a specific type of glutamate receptors (AMPA receptors) to the synaptic region from intracellular storage in vesicles, which initiates LTP and stimulates gene expression in the nucleus, the latter being prerequisite for new synapses.

Presynaptic PAF receptor antagonists can impair memory (Kim and Alger 2004). Studies with PAF receptor knockout mice (genetically manipulated mice that lack an active PAF receptor) had reduced LTP upon stimulation of some (lateral) neural networks, but LTP was not affected in other (medial) neural networks. It was also found that PAF produces an increase in calcium influx into neurons, perhaps via the occupied PAF receptor causing changes in voltage-regulated calcium ion channels in the presynaptic axon terminals (see appendix E).

As mentioned in earlier discussions, there are three different forms of cyclooxygenase (COX) enzymes known, COX-1, COX-2, and COX-3. COX-3 is derived from COX-1 through alternative m-RNA processing (retention of intron-1 of COX-1 mRNA) and seems to be found mostly in the brain (Shaftel et al. 2003). COX-2 has received the most attention of these three isozymes, because it is inducible in many tissues (see chapter 9), but it is constitutively expressed in brain and some other tissues (Kim and Alger 2004). Selective COX-2 inhibitors were developed primarily for treatment of inducible COX-2 involved in inflammation and arthritis.

COX-2 is expressed in certain neurons of the cortex and hippocampus, where it may be playing a role in LTP and memory. Expression of COX-2 increases with increased excitatory (glutamate) synaptic activity and is associated with LTP. COX-2 has been localized in dendritic spines (postsynaptic membranes with a high density of receptors and associated proteins). There is increasing evidence that COX-2 mediates hippocampal LTP, a vital component of memory, through production of prostaglandins and other eicosanoid derivatives (Yang and Chen 2008). Administration of COX-2 inhibitors diminished LTP. This suggests that the ill-fated selective COX-2 inhibitors that increased the risk of blood clots in the heart (myocardial infarction) may have also contributed to memory impairment in the elderly population using these drugs for arthritis. However, there is evidence that nonsteroidal anti-inflammatory drugs (NSAIDs or COX inhibitors) may slow the progression of Alzheimer's disease (see chapter 13). LTP has not been affected by blocking COX-1 in a variety of studies. The suppression of LTP by COX-2 inhibition appears to be elicited within the postsynaptic cell, with decreased excitability and decreased calcium influx. Whether this effect is via actions on the NMDA receptor or other calcium channels is not clear. The actions attributed to COX-2 activity in LTP appear to be mediated primarily through PGE_2 and not PGD_2 or $PGF_{2\alpha}$.

Just as traditional neurotransmitters have multiple receptors, PGE_2 does as well. At least four different types of prostaglandin E receptors have been identified with three different actions elicited in cells where they are located. Two of them (EP2 and EP4) stimulate adenyl cyclase and increase cyclic AMP production, which in turn activates protein kinase A. One (EP3) inhibits adenyl cyclase and suppresses cyclic AMP production. Another (EP1) activates phospholipase C (PLC) to initiate a cascade of events that activates protein kinase C (PKC) and increases the level of calcium in the cell. Calcium increases in this case as a result of inositol trisphosphate (IP_3) released from membrane phospholipids, stimulating calcium channels in intracellular compartments (Cimino et al. 2008). These are the same actions described for the four major epinephrine receptors described in chapter 4.

The different E-type prostaglandin receptors seem to be diversely distributed in pre- and postsynaptic neurons, as well as in astrocytes (Kim and Alger 2004). It is not clear whether calcium influx into the cytoplasm of neurons via NMDA receptors (mentioned earlier) elicits the same or a different response compared with its efflux from intracellular compartments via inositol triphosphate–stimulated calcium channels in organelles. The latter mechanism for calcium release could occur when PGE_2 binds EP1 receptors. However, EP1 receptors are usually located on axon terminals, whereas NMDA receptors for glutamate are generally on dendrites or cell bodies. Therefore, they are in different parts of neurons and may evoke different effects. Different biochemical pathways in different regions of a cell may be affected separately by these two different mechanisms for increasing calcium in the cytoplasm.

COX-2 has been shown to catalyze another biochemical reaction in addition to synthesis of prostaglandins and thromboxanes, namely the conversion of endocannabinoids, 2-AG and/or anandamide, to their respective glycerol and ethanolamide prostaglandin derivatives (Yang et al. 2008). Since COX-2 inhibition can increase release of the inhibitory neurotransmitter GABA in the hippocampus, it is possible this action is mediated through decreased metabolic inactivation of 2-AG and/or anandamide. Consequently, more of these endocannabinoids in the brain would be expected to suppress memory. Since COX-2 is inducible and its levels can increase in response to certain conditions in the brain, this may be another way of regulating levels of endocannabinoids—another level of complexity that will have to be dealt with in future experiments. It will be interesting to see whether aspirin inhibition of COX-2 will alter its enzyme activity with regard to endocannabinoid metabolism, as it did for lipoxin metabolism (see chapter 9).

Conclusions

It is clear that the brain has unique requirements for essential fatty acids, many of which are not easily assessed by carefully controlled clinical studies of the effects of dietary manipulations or supplements. There is no question regarding the requirement for omega-3 PUFAs in the developing brain of humans, although there does seem to be some controversy about whether omega-3 supplements are necessary for normal, healthy infants and children. If infants are being breast-fed and the mother has an adequate supply of omega-3 fatty acids in the diet, there should be no need to provide additional supplements. However, if infants receive formula instead of breast milk, there should be assurance of an adequate (not excessive) supply of omega-3 long-chain polyunsaturated fatty acids (especially DHA) in the formula.

Dietary supplementation with omega-3 fatty acids seems to be a natural recommendation for women from early stages of pregnancy through lactation. There are no known undesirable side effects, other than palatability of certain forms of fish oil supplements (a spoonful of cod liver oil is rarely seen today), provided there are adequate omega-6 fatty acids in the diet. This latter provision would not seem to be a concern to anyone consuming a normal diet (Western or otherwise). Women who consume a reasonable amount of fish or seafood may not need to be concerned about omega-3 supplements. Although the growing abundance of farmed fish may result in lower levels of omega-3 oils, depending on the sources of nutrients for those fish, there are still more omega-3 fatty acids in farmed fish than in most other animal sources.

It seems that health professionals could easily assess whether a woman is getting adequate omega-3 oils in the diet by asking just a few simple questions regarding frequency of fish and seafood consumption. If there is any doubt, there should be no hesitation in recommending omega-3 supplements. Any amount of supplement consumed will be better than none at all in cases where omega-3 intake is low throughout life. It seems that such attention to dietary omega-3 status requires very little in terms of time, effort, and cost but can provide immeasurable benefits in terms of the quality of life and mental health of a future generation.

Although roles of long-chain polyunsaturated fatty acids in learning and memory are only beginning to be worked out, there is much evidence that both omega-3 and omega-6 polyunsaturated fatty acids are essential. The complexities of the brain preclude any proven recommendations for beneficial effects of dietary supplements in general; nevertheless our ancestors' recommendation for a spoonful of cod liver oil every day seems to merit some validity. The following chapters will emphasize the importance of omega-3 oils on many other aspects of the nervous system.

12

Functional Disorders
of the Nervous System

Neurological disorders can be placed into two broad categories, neurodegenerative diseases and functional disorders. The neurodegenerative diseases are characterized by gross loss of specific types of cells within specific tracts of the nervous system, with inflammation as a common thread that ties them together. These include Alzheimer's, Parkinson's, and Huntington's diseases, amyotrophic lateral sclerosis (ALS or Lou Gehrig's disease), and multiple sclerosis (MS). These will be discussed in the next chapter.

The functional disorders are less well characterized in terms of specific neurological deficits but in some cases may be associated with aberrant function of specific types of neurons, while in others the etiology is uncertain. Depression is usually associated with a malfunction of the serotonergic neurotransmitter system because drugs that block the reuptake of serotonin from the synapse are used to alleviate the symptoms of this disease. Similarly, schizophrenia is often associated with the dopaminergic neurotransmitter system because drugs that block dopamine receptors can alleviate some symptoms of this disease. Attention deficit/hyperactivity disorder (ADHD) is treated with drugs that block the reuptake of dopamine and norepinephrine from the synapse, but these drugs (methylphenidate and amphetamine) are generally classified as stimulants, and their efficacy in the treatment of ADHD is puzzling—why would a stimulant be effective in suppressing hyperactivity?

It is tempting to use the known actions of drugs on specific neurotransmitter systems to support theories related to the role of these transmitters in neurological disorders. Indeed, theories of drug action have been extended to suggest theories for the neurochemical dysfunctions, such as schizophrenia being due to overactivity of the dopaminergic system because drugs that block dopamine can alleviate some of the symptoms. However, there are problems in explaining why symptoms of the disorders take weeks to disappear, while the drugs reach their effective levels in the brain within hours or days and should have their effects almost immediately if the theories are correct. Variance between theory of drug action and etiology of disease is especially prevalent in the cases of schizophrenia and depression. There are many unknowns that lie between drugs binding to their known targets in the brain and subsequent changes in mood or behavior. This murky area may be where dietary fats or lipids enter the picture.

Overview of Functional Disorders

The functional disorders, such as ADHD, dyslexia, impulsive disorders, anxiety, schizophrenia, and affective or mood disorders may be a continuum of neurological disorders that have similar origins but affect neurotransmitter systems and consequent behavior in different ways. Studies have shown beneficial effects of omega-3 oil supplementation for some types of each of these disorders, although results are not always consistent. There are many bioactive products produced from long-chain polyunsaturated fatty acids (LCPUFAs), including several prostaglandins, lipoxins, docosanoids, and endocannabinoids, that play important roles in the brain. Profound differences between omega-3 and omega-6 LCPUFAs and their bioactive metabolites support arguments that large shifts in the ratio of omega-6 to omega-3 fatty acids in modern Western diets may be promoting these disorders in susceptible individuals (Simopoulos 2002a).

Environmental factors can play a role in the occurrence of neurological disorders, particularly schizophrenia and affective disorders, although environment causes only a minor increase in risk. In the cases of schizophrenia and affective disorders, people born in winter or spring months have a greater risk of developing those disorders, which has roused speculation that pathogens such as viruses may be contributing to developmental abnormalities (Strange 1992, 279). Premature births and birth complications have also been implicated. In view of the acute vulnerability of the developing brain to environmental factors in the late stages of pregnancy and early infancy, it should not be surprising that nutritional deficiencies, infections, and various traumas have some influence on neurological development. This is the time when neurons are proliferating rapidly and synaptic connections are being orchestrated. Events or substances that interfere with neuronal wiring at this fragile stage could have long-term consequences that may not appear immediately.

There could also be hereditary influences on manifestations of these illnesses. Genetic variants are known or suspected to increase the risk of several disorders. Many enzymes (products of gene expression) have been implicated in these disorders, including several involved in metabolism of bioactive lipids (such as phospholipase A_2 and desaturases). It is important to keep in mind that genetic variation may be only part of the underlying causes of these diseases and the relative success or failure in their treatment.

The amine neurotransmitters, particularly dopamine, norepinephrine, and serotonin, are implicated in these disorders to varying degrees. However, these neurotransmitters are the easiest to monitor because they can be measured easily in cerebrospinal fluid or their metabolites can easily be measured in urine. This is like the drunk who is looking for his lost keys under the streetlight because that is the only place where he can see in the darkness. Implications can be made from differences in levels of these transmitters or their metabolites, but it is not possible to state with certainty that a particular neurotransmitter system is the only one involved, simply because that was the only one for which differences could be measured. Peptide neurotransmitters are seldom measured in neurological disorders, and their actions in the brain are not well understood. Each neurotransmitter may also affect many other transmitter systems through signals to other neurons, and ultimately there may be influences on neuronal outgrowth to modify synaptic connections (see chapter 11). We are currently quite far from understanding the etiology of these disorders.

LCPUFAs, particularly arachidonic acid and docosahexaenoic acid (DHA), clearly have influences on development of the nervous system in the fetus and newborn, as discussed in the previous chapter. They probably play an important role in learning and memory throughout life. Since LCPUFAs and their precursors can be stored in lipid deposits when they have been consumed in excess of their needs, it becomes difficult to assess their role in a wide range of neurological disorders. It is also not easy to establish whether an essential fatty acid deficiency exists in human subjects, because of numerous interactions of various nutrients and the fact that most human diets have an abundance of essential fatty acids, especially linoleic acid and, to a lesser extent, linolenic acid. Because of the dramatic increase in the ratio of omega-6 to omega-3 fatty acids in modern Western diets and the well-established importance of omega-3 fatty acids, particularly DHA, in neurological function, recent research in nutritional neuroscience has focused on dietary omega-3 fatty acids and their effects on a wide range of neurological disorders.

The research journal *Prostaglandins, Leukotrienes and Essential Fatty Acids* devoted a special double issue in 2000 to a multidisciplinary workshop for researchers and clinicians to explore the interactions of essential fatty acids and phospholipid metabolism with developmental neurological disorders, such as attention deficit/hyperactivity disorder (ADHD), dyslexia, dyspraxia (developmental coordination disorder), and autism. These are among the most common neurological disorders in childhood, although their identification and management are still controversial. In April 2006 the *International Review of Psychiatry* devoted an issue to the role of essential fatty acids in neurological and psychiatric disorders. In addition, there has been much interest in exploring the effects of dietary essential fatty acid supplements on depression and schizophrenia, as well as cognitive decline and other neurological deficits associated with aging.

Although some interesting relationships have been reported, most researchers would not agree that essential fatty acid supplements are going to be the answer to alleviating many of these neurological disorders. Their etiology is complex and multifactorial, and little progress has been made in identifying the biochemical parameters responsible for predisposing individuals to these disorders. However, a closer look at the results of some of these studies begins to reveal a general concurrence that omega-3 fatty acids are playing a critical role in ubiquitous neurological functions. It would behoove us to heed these cues and ensure that everyone is getting the necessary amounts of omega-3 fatty acids throughout life. Megaquantities are not necessary, but adequate amounts over the course of a lifetime probably are needed to help stave off a wide range of deficits. Indeed, in many studies moderate amounts of supplements were more beneficial than larger amounts.

The following discussion of important findings related to essential fatty acid supplements in wide-ranging neurological disorders should lead the reader to conclude that a good balance of omega-3 to omega-6 fatty acids in the diet can result in a healthier mind and body. The previous chapter gave a general overview of the role of the omega-3 fatty acids, particularly DHA, in early development of the nervous system and its continuing importance throughout life for learning, memory, and normal cognitive function.

Attention Deficit/Hyperactivity Disorder

Children having problems paying attention, listening to instructions, and completing tasks may be diagnosed as having attention deficit/hyperactivity disorder (ADHD). Such children may blurt out answers, interrupt others, and generally disrupt class. Hyperactive children have more auditory, visual, language, reading, and learning problems than their nonhyperactive counterparts (Mitchell et al. 1987). They also tend to have lower birth weights (3.058 versus 3.410 kg) and have significantly more frequent colds, polydypsia (excessive thirst), polyuria (frequent urination), and serious illnesses or accidents. In some studies they were found to have greater frequency of asthma, eczema, and allergies (Colquhoun and Bunday 1981).

Hyperactive children had significantly lower levels of LCPUFAs (DHA, arachidonic acid, and dihomo-gamma-linolenic acid) in their blood relative to a control group (Mitchell et al. 1987). These findings were the basis for numerous subsequent studies to determine whether essential fatty acid supplements would have any impact on behavior and learning in ADHD children. The fact that they had significantly lower birth weights suggests these children came into this world with a DHA deficiency, since maternal DHA status has been correlated with birth weight (as discussed in the previous chapter).

Several other studies have shown that children with ADHD have lower levels of LCPUFAs in their blood (plasma phospholipids and red blood cells), particularly the omega-3 type (Burgess et al. 2000). John Burgess's group at Purdue University studied children with ADHD that also exhibited classic signs of essential fatty acid deficiency, such as skin conditions and excessive thirst (Antalis et al. 2006). They also found that college students with ADHD had a tendency to have skin and thirst symptoms indicative of essential fatty acid deficiency and had significantly lower levels of omega-3 fatty acids in blood phospholipids and red blood cells compared with a control group of college students. Dietary fatty acid intakes of the ADHD group did not differ significantly from those of the control group, which led the authors to speculate that the differences in fatty acid profiles of blood components might be due to a metabolic insufficiency (genetic in origin).

The fact that thirst and frequent urination are associated with ADHD might suggest a vasopressin (antidiuretic hormone) insufficiency in this neurological disorder. It was mentioned in the previous chapter that vasopressin has been shown to affect memory, although its role in the memory process is not understood. Vasopressin is a nonapeptide (protein with nine amino acids) with a disulfide connecting two cysteine residues. As a peptide hormone/transmitter it is not as easily studied as the classical amine transmitters. There does not appear to have been any study to determine whether there is a correlation between severity of ADHD symptoms and vasopressin levels in the blood. It is also possible that a genetic mutation in either the gene for vasopressin or the genes for vasopressin receptors is the basis for some cases of ADHD and related learning disorders. Along these same lines, but from a different angle, several studies have found that vasopressin binding to its V1a receptor can stimulate aggressive behavior, and an experimental drug that blocks this receptor was shown to suppress aggression in animals (Ferris et al. 2006). Since vasopressin is a peptide hormone, it would not be effective if taken orally, because it would be broken down to amino acids in the gastrointestinal tract. Vasopressin and bioactive analogs of this peptide hormone are available for inhalation therapy.

An early study found evidence of zinc deficiency in ADHD children (Brophy 1986). Zinc is a cofactor in the delta-6-desaturase enzyme involved in the conversion of eighteen-carbon essential fatty acids (linoleic and linolenic acids) to LCPUFAs (see figures 2.4 and 3.1), so either a zinc deficiency or a genetic defect in this enzyme could explain lower levels of LCPUFAs in these individuals when they are consuming a diet with normal levels of essential fatty acids (linoleic and linolenic acids). A study of the effect of zinc supplements on ADHD in adolescents showed significant improvement in some of the symptoms of ADHD (hyperactivity, impulsivity, and impaired socialization) but not attention deficit (Bilici et al. 2004). Perhaps increased levels of LCPUFAs (DHA and arachidonate) would be beneficial in the diet along with zinc supplements.

Indicators of elevated oxidative stress were not found in the ADHD group, which might have led to a faster oxidative degradation of omega-3 fatty acids and accounted for the lower level of these fatty acids (Mitchell et al. 1987). Another group of researchers found that children with ADHD exhaled higher levels of ethane, which is a marker for oxidative breakdown of omega-3 fatty acids in the body and would suggest greater oxidative stress (Ross et al. 2003). However, one must use caution in interpreting the ethane exhalation results, since the site of oxidation of the omega-3 fatty acids is not known when monitoring breath and could be intestinal in origin; intestinal flora could be causing the production of ethane from breakdown of omega-3 fatty acids in the diet.

Although lower levels of essential fatty acids in red blood cell membranes are consistently found in ADHD children as well as adults, the biochemical underpinnings of these observations are not clear. Essential fatty acid dietary supplements are found to improve the levels of these fatty acids in the blood, but there have not been uniform findings for behavioral improvements regarding the symptoms of ADHD (Weber and Newmark 2007). There have been hints that some behavioral parameters may have improved in some cases, yet overall there is not a consistent pattern. A metabolic insufficiency (lack of an enzyme due to a genetic mutation) could give rise to consistently lower levels of essential fatty acids in blood, although specific enzymes or genes that would be responsible for such metabolic defects have not been identified. Increased oxidation of essential fatty acids also does not seem to be the cause of differences in blood parameters. So, although differences in blood levels of essential fatty acids are consistent, the cause is not clear. As with many disorders, there could be any number of metabolic and/or physiological abnormalities that are contributing.

Brian Hallahan and Malcolm Garland (Hallahan and Garland 2004) have discussed ADHD, as well as some other impulsive disorders such as hostility, aggression, suicide, and deliberate self-harm, and indicated that all have been shown to be associated with low blood levels of some omega-3 fatty acids, either DHA or EPA. The few studies that have been reported for omega-3 supplements as a treatment for aggression have indicated favorable results (no mention of vasopressin involvement). Hallahan and Garland indicate that the serotonergic system is another factor involved in these disorders. Several studies have found lower than normal levels of the serotonin metabolite 5-hydroxyindoleacetic acid (5-HIAA) in cerebrospinal fluid or blood of affected individuals.

Animal studies have shown that DHA and arachidonic acid supplementation can increase levels of serotonin as well as dopamine in the brain (de la Presa Owens and Innis 1999, 2000). It is generally believed that low levels of 5-HIAA in cerebrospinal fluid or urine are due to either low serotonergic neuronal activity or less serotonin being released when a neuron is firing (owing to lower levels of this transmitter in the nerve terminal). The increased levels of neurotransmitter with essential fatty acid supplements may be due to a more efficient serotonergic system when DHA and/or arachidonate have been restored to their optimum levels in the brain. Because DHA is vital to neurological membranes, particularly in the synaptic region, DHA supplements could improve operation of receptors. They could also induce more efficient reuptake of transmitter from synapses via transporters in the synaptic membrane.

Dyslexia and Dyspraxia

Dyslexic children with signs of essential fatty acid deficiency showed poorer reading skills and auditory working memory, as well as lower general ability, than children with lesser clinical signs of essential fatty acid deficiencies (Richardson et al. 2000). These relationships were stronger in females, which is interesting when considering that dyslexia is much more common in males than females. Signs of essential fatty acid deficiency in dyslexic adult males were also strongly correlated with severity of both visual and auditory dyslexic symptoms, as well as linguistic and motor problems, lending further support to the hypothesis that there may be abnormalities in essential fatty acid and/or phospholipid metabolism in dyslexia (Taylor et al. 2000). A magnetic resonance imaging (MRI) study of dyslexic versus control adults showed that there were differences in the ratios of phosphate monoesters relative to diesters, reflecting differences in membrane phospholipid metabolism in these two groups (Richardson et al. 1997).

There have been few studies of the effects of essential fatty acid supplements on dyslexia and dyspraxia. A small study found that dyslexic children have poorer dark adaptation than normal control children and that DHA supplements improved dark adaptation in the dyslexic children but had no significant effect on normal controls (Stordy 2000). It is tempting to propose that DHA plays some role in dark adaptation, in view of its high levels in the retina, and this condition may be worth evaluating as a counterpart to learning problems and assessing whether DHA deficiency may be contributing.

Vitamin A may be a factor in dark adaptation, but there was no evidence of vitamin A deficiency among subjects in that study. One control subject in the study was a vegetarian with poor dark adaptation that improved with DHA supplements. The report also showed that children with dyspraxia exhibited improved motor skills after four months of high DHA and antioxidant supplementation. Although this is one report with a small cohort, the results suggest that omega-3 supplements could be beneficial for these disorders.

Epilepsy

Epilepsy is characterized by hyperexcitability (overexcitation) of a collection of neurons in the brain as a result of either deficient inhibitory brain activity (too little GABAergic activity) or excessive activity of the excitatory glutamate system. The location and extent of the excessive excitatory activity will determine whether the patient experiences partial or generalized seizures. The type of seizure is measured by electroencephalography (EEG).

Hippocrates recommended fasting for epileptic patients. A resurgence in the recommendation for fasting occurred in France around 1911. In the 1920s a high-fat, low-carbohydrate diet was instituted for epilepsy in the United States, with the rationale that this diet would mimic the ketosis induced by fasting or caloric restriction (Geyelin 1921). The dietary approach to treating epilepsy fell out of favor in the late 1930s with the introduction of anticonvulsant drugs, beginning with diphenylhydantoin. Although the ketogenic diet remained a curiosity on the fringes of mainstream medicine for the next sixty years, it experienced a second resurgence in the mid-1990s when a young boy with epilepsy did not respond to anticonvulsant drugs but was successfully treated with a ketogenic diet (Huffman and Kossoff 2006). This led to creation of the Charlie Foundation, which promotes the ketogenic diet for treatment of epilepsy and has stimulated scientific research on the physiological basis for the mechanism of action of ketosis on this neurological disorder.

The ketogenic diet, as originally proposed in 1921, recommended a high-fat (75 percent of calories), low-carbohydrate (10 percent), and adequate-protein (10 to 15 percent) diet, or a ratio between three to one and four to one for fat relative to carbohydrate and protein. The Atkins diet, which is high fat, high protein, and low carbohydrate, has also proven to be effective, often with some minor modifications. Ketones, such as acetone and acetoacetate, are two major ketone bodies circulating in the blood during ketosis that have been found to suppress seizures in certain animal models. A third ketone body, beta-hydroxybutyrate, is not actually a ketone but a metabolic product of acetoacetate. It is in relatively high concentrations in the circulation during ketosis but does not seem to be as effective as the ketones in suppressing seizures (Bough and Rho 2007).

It is not certain which components of the ketogenic diet are responsible for its anticonvulsant effects. Originally it was thought that ketone bodies formed from metabolism of saturated fatty acids under conditions of low blood sugar were primarily responsible for the anticonvulsant effect. This was supported by the efficacy of acetone and acetoacetate in controlling seizures. The ketogenic diet is most effective against seizures induced by GABA antagonists, which block receptors for inhibitory signals. Ketogenic diets are less effective against seizures induced by glutamate agonists that activate receptors for excitation. It has been suggested that ketosis can shift metabolic pathways that favor synthesis of GABA (Yudkoff et al. 2005).

These are just some of the ways the ketogenic high-fat, low-carbohydrate diets may elicit their beneficial effects on epileptic seizures. It has been found that the ketogenic diets can offer as much benefit or more in controlling seizures as drugs (Huffman and Kossoff 2006). Studies of the efficacy of these diets found that about one-third of patients have greater than 90 percent reduction in seizures and about one-third have between 50 and 90 percent reduction, while about one-third have less than 50 percent reduction in seizures (Thiele 2003).

Two recent reviews (Huffman and Kossoff 2006; Bough and Rho 2007) give the latest views of how the high-fat, low-carbohydrate diet may work to control epileptic seizures but indicate there is still much to learn in order to know just what aspects of the diet are most efficacious and how to minimize side effects.

Recent interest in omega-3 fatty acids for various neurological disorders led to studies that showed marked reduction in seizure frequency and intensity when omega-3 supplements were provided in combination with dietary restriction regarding carbohydrates (Schlanger, Shinitzky, and Yam 2002). Omega-3 PUFAs could control seizures in several ways, including effects on sodium and calcium channels or pumps that would suppress excessive excitation (Vreugdenhil et al. 1996). Increased ratios of omega-3 to omega-6 PUFAs in plasma membranes can increase activity of sodium pumps (Wu et al. 2004), which would restore neurons to their repolarized state more quickly. Greater resting potential (voltage) in neurons, as a result of a more active sodium-potassium pump, would shift those neurons further away from the threshold potential that triggers firing (see appendix E for descriptions of these neurological processes).

As mentioned earlier, PUFAs influence expression of specific genes (via the peroxisome proliferator-activated receptor alpha, PPARα), which in turn can increase efficiency of energy metabolism in cells. Reduced expression of mitochondrial uncoupling proteins would decrease production of reactive oxygen species (Garlid, Jaburek, and Jezek 2001). Free radicals and reactive oxygen species, through lipid peroxidation, make membranes leaky and more susceptible to depolarization (excitation). The role of omega-3 fatty acids, DHA in particular, in forming antioxidant plasmalogens should also be considered as a possibility for protection.

Depression and Affective Disorders

In the early 1980s it was postulated that eicosanoids may play a role in depression (Horrobin and Manku 1980). By the late 1980s results of clinical trials using diet, behavior modification techniques, and drugs to lower cholesterol with the aim of decreasing risk of heart disease showed a most unusual effect—increased incidence of suicide, aggression, and violent deaths (Lipid Research Clinics Study Group 1984). Joseph Hibbeln and Norman Salem hypothesized that the increased incidence of depression and suicide was likely due to the dietary recommendations that were used to achieve lower blood cholesterol levels, namely substituting polyunsaturated vegetable oils for saturated fats in the diet (Hibbeln and Salem 1995). Such practices were inadvertently increasing the ratio of omega-6 to omega-3 PUFAs consumed.

Hibbeln and Salem pointed to increasing rates of depression in North America throughout the twentieth century, as documented in the Epidemiological Catchment Area survey (Klerman and Weissman 1989). They postulated that dramatic increases in vulnerability to depression were due to dietary shifts toward omega-6 polyunsaturated oils from a few plant sources. The ratio of omega-6 to omega-3 PUFA consumption went from about three to one at the beginning of the century to perhaps more than twenty to one by the end of the century.

The increase in depression parallels the increase in heart disease during this time period. It is tempting to conjecture that dietary patterns were responsible for both. Depression has been suggested as a risk factor for heart disease, although it is not widely accepted nor as easily assessed as other, more common risk factors (Pozuelo et al. 2009). Because DHA, the dominant omega-3 PUFA in the brain, is important for many neurological functions, this large change in the ratio of dietary PUFAs could be responsible for a number of neurologic disorders.

Numerous studies have shown that populations that eat more fish have much-reduced incidence of depression compared with populations that eat little seafood. Perhaps this effect was most dramatic in a Finnish study in which low cholesterol levels were associated with lower mortality from accidents and violence in coastal Finland—where fish is prevalent in the diet. When the researchers compared an inland Finnish population, where fish is less common in the diet, there was no association between cholesterol levels and accidental or violent deaths (Pekkanen et al. 1989). In the coastal regions, those who ate fish would have had lower levels of cholesterol and higher levels of omega-3 fatty acids than those who did not eat fish.

There has been a glut of reports since the mid-1990s that show an inverse relationship between levels of omega-3 PUFAs in red blood cell membranes (a good marker for omega-3 fatty acid status) and the severity of symptoms in depressed individuals (Parker et al. 2006). A recent analysis of published trials (Appleton et al. 2006) of effects of omega-3 supplements on depression indicates a complex picture. Most studies in the analysis show some benefit with omega-3 supplements, although some do not show a benefit. But none of the studies show any significant adverse effects of omega-3 supplementation.

What is perhaps the most interesting is that some trials using the highest levels of omega-3 supplements show the least beneficial effect, whereas moderate levels of omega-3 fatty acid supplements generally show greater beneficial effects. The studies are complicated by the types of mood disorders that were selected for study and the type of supplements given (DHA, EPA, or a combination of these two). Some studies looked at bipolar disorders, others monopolar depression, and some postpartum depression.

There is a decrease in prevalence of postpartum depression in populations with high levels of seafood consumption and high levels of DHA in breast milk (Hibbeln 2002). A small short-term (two weeks prior to birth) study of omega-3 supplementation and another of omega-3 supplementation for four months during breast-feeding did not show beneficial effects on postpartum depression. The timing and duration of these studies do not permit firm conclusions to be drawn regarding lack of effect. The fetus is going to be drawing much of whatever DHA is available from the mother in the last two weeks before birth, so supplements given to the mother during lactation and breast-feeding would probably go directly to the mother's milk, with relatively little remaining for general distribution in her body. A more logical approach to the use of omega-3 supplements for pregnant women is to begin the supplements as early as possible in the pregnancy, or even before pregnancy, and continue their use throughout lactation and breast-feeding. This will provide both mother and child with optimum benefits of ample omega-3 fatty acids, hopefully circumventing any complications for the fetus or infant and minimizing risk of postpartum depression in the mother. In addition, the improved DHA status of the child could have an impact on his/her intellectual and psychological development later in life.

One report describes a single male patient with treatment-resistant, severe, and unremitting depression over a period of seven years who received an eicosapentaenoic acid (EPA) supplement along with conventional antidepressant treatment for a nine-month period. He exhibited dramatic and sustained clinical improvement in all symptoms of depression, including severe suicidal ideation, within the first month. Symptoms of social phobia also improved. MRI at the beginning and end of the nine-month supplementation period showed structural brain changes, particularly a reduction in volume of the lateral ventrical (Puri et al. 2001). The same group of researchers has published similar reports of individual cases showing clinical improvements with omega-3 supplements.

The lack of benefit with large-dose supplements in some studies of affective disorders may be due to high levels of EPA relative to DHA and arachidonic acid. One could speculate that EPA would compete with arachidonic acid for synthesis of bioactive eicosanoids or inhibit the utilization of arachidonate. Some trials using ethyl-EPA, an ester of EPA that gets hydrolyzed in the body to give EPA, showed beneficial effects when given at low to moderate levels. Some trials with a combination of EPA and DHA were also effective, while trials with only DHA supplements did not show beneficial effects. Perhaps the utility of EPA is due to its inhibition of omega-6 fatty acid metabolism. There are still many questions regarding the most efficacious omega-3 supplements and dosage in treating mood disorders, but it appears they can provide some benefits in many, but perhaps not all, cases when used in moderation.

It is not clear which omega-3 fatty acid, EPA or DHA, is the ultimate agent in the effect of these fatty acids on mood disorders. Drugs for treating mood disorders, namely lithium, carbamazepine, and valproate, decrease the turnover of arachidonic acid and decrease the production of PGE_2 (Sinclair et al. 2007). It is possible that EPA could show beneficial effects through metabolism to different eicosanoids than are formed from this fatty acid. Alternatively, EPA could merely be inhibiting arachidonic acid metabolism to eicosanoids. Although EPA could exert its effects by being metabolized to DHA, the lack of efficacy by DHA alone in some trials seems to indicate that EPA is playing a role in its own right. DHA is known to have many effects in the brain, including effects on transmitter synthesis, receptor binding, signal transduction, transporters for reuptake of transmitters, and activities of many membrane-associated enzymes and proteins.

Schizophrenia and Related Psychoses

Schizophrenia is a range of psychological disorders that generally strike younger adults (and older adolescents) with so-called positive symptoms, while many cases exhibit chronic illness displaying predominantly negative symptoms. The broad range of symptoms include disordered thinking, delusions, perceptual disturbances, mood disorders, and sporadic movements, which are all classified as positive symptoms. Negative symptoms include social withdrawal, diminished speech, reduced motor function, and loss of emotion, which are more frequently associated with long-term illness. The acute positive symptoms usually respond to treatment with antipsychotic drugs, although response to drug treatment is quite variable—some patients recover completely; others may never recover and continue to get worse. Large numbers of patients fall between these extremes, relying on management of recurring episodes with antipsychotic drug therapy.

In the late 1980s an article postulated that there was a relationship between the incidence of schizophrenia and the national dietary intake of fat. The relationship was even stronger when the ratio of saturated fats to polyunsaturated oils was used as a predictor (Christensen and Christensen 1988). Although this report focused on fats, dietary preferences are complex. Higher saturated fat consumption will often correlate with higher refined (or processed) sugar consumption. Multiple regression models proved sensitive to the way independent variables were entered. Sugar proved to be the dominant predictor of poor outcome for schizophrenia (Peet 2004). Saturated fat (from butter or ghee) gave a positive correlation for poor outcome, albeit weaker than the correlation with sugar. Fish and seafood showed a negative correlation. Fish consumption was associated with a favorable outcome. Another report indicated lower lifetime prevalence of schizophrenia and better outcomes in Asian as opposed to non-Asian countries (Goldner et al. 2002). Many factors could be contributing to this latter statistic, dietary fats and sugar being only some of them.

Early supplementation studies in the 1980s showed that evening primrose oil (an omega-6 oil) provided no benefits for psychoses (Vaddadi et al. 1986), whereas linseed oil (high in omega-3) supplements showed significant improvements (Rudin 1981). Red blood cell membranes were found to be deficient in essential fatty acids, particularly arachidonate and DHA, just as they are for other neurological disorders discussed above. Several clinics corroborated the lower levels of essential fatty acids in red blood cell membranes of schizophrenic patients versus controls, which led to numerous studies of essential fatty acid supplementation (Peet 2004; Berger, Smesny, and Amminger 2006). Omega-3 oils, particularly EPA, were generally more effective than omega-6 supplements. Several groups have now reported various degrees of improvement in symptoms of schizophrenia when omega-3 supplements were used in combination with ongoing antipsychotic medications. EPA supplements seem to be more effective with clozapine than with other drugs.

It is curious that one trial that did not show improvements with EPA supplements was in the United States, a result that may be related to high-fat and high-sugar consumption. Restricting intake of sugar and omega-6 oils during treatment with EPA or fish oil supplements would likely give more promising outcomes for schizophrenia patients. Such strategies need to be explored when future dietary studies are designed.

Malcolm Peet (Peet 2004) argues that high-sugar, high-fat diets have been shown to reduce hippocampal brain-derived neurotrophic factor (BDNF) in rats. BDNF is important in dendritic growth and maintenance, which plays an important role not only in learning but in synaptic communications in general. Reduced BDNF expression has also been observed in the prefrontal cortex of schizophrenia patients, lending further support to this explanation regarding the mechanism.

Autism

Autism is a severe developmental disorder that may originate at birth or in the first few years of life. It is characterized by social and behavioral problems. There is much evidence for real and dramatic increases in the incidence of this disease from the early 1990s into the first decade of the twenty-first century. Some argue that the increase may be due to changes in diagnosis. However, there has been more than a doubling, and perhaps as much as quadrupling, of incidences since the 1980s in most developed countries. The risk of autism is associated with a variety of factors, such as older parental age and low birth weight—the latter again suggesting a possible DHA deficiency, although a firm relationship has not been established.

There has been much controversy regarding potential environmental causes, particularly vaccines, but the scientific evidence is lacking. In fact, this hypothesis was disproved in the case of measles/mumps/rubella (MMR) vaccine (Honda, Shimizu, and Rutter 2005). Until the causes of this disease are better understood, it will be difficult to make recommendations regarding dietary fatty acids or supplements. Attention should be given to proper omega-3 dietary fatty acid intake, as for many other disorders.

There are several similarities between autism and schizophrenia, with the major difference being age of onset. It would seem valuable to determine whether there is any association between this disorder and exposure to sugars, since there does seem to be some relationship between sugars and schizophrenia. Although high-fructose corn syrup might arise as a suspected sugar in terms of an association with the increase in autism, this sweetener is not commonly used outside the United States. It would be important to monitor sugar intake during fetal and postnatal development.

Conclusions

It should be clear that omega-3 fatty acids play important roles in the brain, and there is much evidence that omega-3 essential fatty acid insufficiency could be contributing to many neurological disorders. Although trials of dietary supplements have given mixed results in many studies, there have not been reports of omega-3 supplements exacerbating symptoms of any of these disorders. However, keep in mind that high levels of omega-3 supplements proved to be less effective than moderate doses in some trials for affective disorders. Similar observations with other disorders were not seen, perhaps because large ranges of dosage were not tested. The lack of beneficial effects of omega-3 supplements in some trials could be due to inattention to limiting the intake of omega-6 oils in many trials.

Considering the abundance and vital roles of DHA in neurological development from the fetal stage through infancy, childhood, and beyond, there should be no hesitation in recommending that everyone get adequate amounts of omega-3 fatty acids throughout life. Excessive doses from dietary supplements should be avoided, particularly if there is no reason to suspect DHA deficiency. If a person's diet and lifestyle are such that DHA deficiency is likely, analysis of fatty acid profiles in blood lipids and red blood cell membranes could confirm an essential fatty acid deficiency. Even without substantiation of deficiency, omega-3 supplements in moderation, with attention to limiting omega-6 intake, would be a reasonable approach that people can take without risk of side effects.

Most of the attention paid to dietary aspects of neurological disorders has been focused on fats, but dietary sugars should not be ignored. Sugar showed a stronger correlation than fats with poor outcome for schizophrenia. In view of the rising incidence of type 2 diabetes and metabolic syndrome that can be traced to high sugar in the diet, particularly high-fructose corn syrup, more attention should be given to the possibility that fructose may be contributing to some these neurological disorders. Most of the fructose absorbed from the diet is handled by the liver and converted into fats, but having larger amounts in the diet would just mean there would be more in the circulation at any given time. Little is known about the fate of fructose in the brain, but its inordinate chemical reactivity and propensity for generating free radicals and other reactive products should make it a prime suspect in many neurological disorders. This is certainly an area that demands greater attention in the future.

13

Neurodegenerative Diseases

The neurodegenerative diseases seem to have in common a free-radical-mediated destruction of cellular components, particularly cell membranes. All seem to have inflammation involved. Apoptosis is also prevalent in most of these diseases, but this programmed or systematic suicide of cells appears to be a mechanism for preserving the organism when there is damage to specific cells. The cause of damage is not known for certain in these diseases, and may be multifactorial. In some cases, genetics plays a major role (as in multiple sclerosis and Huntington's disease), whereas in others the genetic link seems to be a minor risk factor and environmental factors are perhaps making more significant contributions.

Multiple Sclerosis

Multiple sclerosis (MS) is one of the most common autoimmune neurodegenerative disorders. T lymphocytes cross the blood-brain barrier and attack oligodendrocytes, destroying the myelin sheath that surrounds long axonal projections in the central nervous system. The destruction of myelin and its accompanying inflammation disrupt nerve impulses, all of which become most prevalent during a relapse. There may be some repair of damaged tissue as well as suppression of inflammation during the remission phase. Over time, however, the damage that accrues during relapses may become extensive and result in considerable loss of function.

The overt symptoms of MS can vary widely from one individual to another. There is generally muscle weakness, which leads to decreased mobility and often results in bladder instability. Difficulty in chewing and swallowing can lead to malnutrition. There is much anecdotal evidence that diet can influence the course of MS, although reviews of dietary intervention trials to manage the progression of MS have been discouraging.

Epidemiology studies indicate that countries with high saturated fat intake have higher incidence of MS than those with higher PUFA intake (Payne 2001), although such gross associations need more careful scrutiny. The rate of MS in Switzerland is inversely associated with altitude of residence, whereas in Norway there is a higher prevalence of MS inland than along the coast (Hayes, Cantorna, and DeLuca 1997). The Norwegian phenomenon has been used to argue that omega-3 fish oils may be beneficial in suppressing MS. The Swiss phenomenon has been explained by higher levels of vitamin D (the sunshine vitamin) in the population at higher altitudes as a result of greater UV light intensity. The latter explanation is further supported by the rarity of MS in the tropics and the dramatic increase in incidence with higher latitude in both Northern and Southern hemispheres.

One study found a large (38 percent) increase in lipid peroxidation and other markers of oxidative stress in MS patients versus controls (Karg et al. 1999). Lipid peroxides are generally elevated when increased immune system activity and inflammation are present. Antioxidant vitamins are often recommended for MS, although clinical trials for such intervention have given mixed results. There is a sound basis for omega-3, particularly DHA, supplements. It was mentioned (in chapter 11) that DHA supplements increased levels of plasmalogens, which have antioxidant properties and are normally at high levels in myelin.

The arguments regarding vitamin D suggest that calcium is important, along with vitamin D. However, vitamin D could be affecting MS in ways that do not involve calcium. In view of the limited information currently available, it seems that adequate amounts of omega-3 oils, vitamins, and minerals, particularly vitamin D and calcium, may be a good approach to slow the progress of this devastating disease.

Amyotrophic Lateral Sclerosis

Amyotrophic lateral sclerosis (ALS) is characterized by progressive loss of motor neurons, particularly in the spinal cord. Patients typically succumb to the disease within five years of onset. As with some other neurodegenerative diseases, there is a small (5 to 10 percent) chance of it being genetic; the majority of ALS cases are considered to be idiopathic. Hereditary ALS has been associated with several genes. The gene for copper-zinc superoxide dismutase (SOD1) on chromosome 21 has received the most attention, but only about one in every five or six cases of familial ALS is associated with mutations in the SOD1 gene.

A transgenic mouse model has been developed using a mutant SOD1 gene that mimics several characteristics of the human disease. This lends support to the hypothesis that free radicals and reactive oxygen species are important components of the neurodegenerative process, and may be generated by an inflammatory immune response (Emerit, Edeas, and Bricaire 2004). Increased levels of proinflammatory substances have been observed in ALS, as well as activation of microglia (Moisse and Strong 2006). There also appears to be elevated levels of the excitatory neurotransmitter glutamate in the cerebrospinal fluid of ALS patients. It is believed that glutamate may be neurotoxic through overexcitation of neurons, leading to neuronal death (Consilvio, Vincent, and Feldman 2004).

COX-2 (the inducible cyclooxygenase enzyme for prostaglandin synthesis) was increased in the spinal cords and PGE_2 levels were higher in the cerebrospinal fluid of ALS patients (Minghetti 2004). There is evidence that COX-2 may take part in excitotoxic damage caused by elevated levels of glutamate. Treatment of SOD1 transgenic mice with a COX-2 inhibitor delayed the onset of disease and prolonged the survival of these mice (Drachman et al. 2002), which would indicate that nonsteroidal anti-inflammatory drugs (NSAIDs) could have a therapeutic effect on ALS patients. However, COX-2 inhibitor was administered prior to the onset of disease symptoms in the mice, and the benefits of drug interaction after symptoms appear have not been investigated. Nitric oxide (NO) synthase is also increased in the transgenic mouse model of ALS. NO is a free radical and may augment further generation of damaging free radicals, in addition to its signaling properties.

Several studies looked at markers for oxidative stress and antioxidant status in ALS patients. Blood from patients with sporadic (idiopathic) ALS had elevated levels of products of lipid peroxidation, although levels of antioxidants (vitamin E, beta-carotene, glutathione, and ubiquinone) and the activity of red blood cell SOD1 were not significantly different from those of healthy control subjects (Oteiza et al. 1997). Another study found decreased levels of erythrocyte SOD1 and glutathione peroxidase in sporadic ALS patients compared with controls, and treatment with a multiple antioxidant supplement improved the glutathione peroxidase activity and vitamin E status but did not affect SOD1 (Apostolski et al. 1998).

There is much evidence for mitochondrial dysfunction in ALS, with some speculation that mutated SOD1 may become associated with mitochondria. It is interesting that mitochondrial dysfunction is not restricted to motor neurons but may be a systemic defect in which motor neurons and skeletal muscle are just more susceptible to the effects. There is impaired mitochondrial respiration and overexpression of an uncoupling protein in mitochondria. The uncoupling protein causes increased energy loss as heat by uncoupling respiration from production of adenosine triphosphate (ATP) (Dupuis et al. 2004). The mitochondria are also involved in sending signals to cells for their programmed demise (apoptosis). The mechanisms for various functional abnormalities in ALS patients, or in animal models, are still not understood.

Patients with ALS exhibit a generalized hypermetabolic state, which is likely due to mitochondrial dysfunction, although alternative causes cannot be ruled out. ALS patients tend to lose weight with onset of the disease, although difficulty in swallowing in advanced stages of the disease accounts for much of the malnutrition. High-fat diets have been found to extend the lifetime and prevent loss of motor neurons in the transgenic mouse model for ALS. Transgenic mice fed a ketogenic (high-fat, low-carbohydrate) diet had better motor performance, longer life span, and weighed more than their normal-fed controls (Zhao et al. 2006). The ketone bodies, as well as fatty acids, would be handled differently than carbohydrates by mitochondria.

Further support for the ketogenic diet comes from an epidemioligic study of risk for ALS in Japan (Okamoto et al. 2007). Using food-frequency questionnaires, these investigators found a significantly higher risk associated with a high intake of carbohydrate (odds ratio = 2.14), whereas risk was significantly lower with a high fat intake. Saturated fat gave the lowest odds ratio (0.30), while polyunsaturated fatty acids gave a less favorable odds ratio (0.58). As mentioned in chapter 12, the ketogenic diet has been used with some success in the treatment of epilepsy.

Huntington's Disease

Huntington's disease is named for George Huntington, who studied the incidence of chorea (uncontrolled movements) in the late 1800s. It is a rare (less than 0.01 percent of the population), inherited disease with an average age of onset between thirty-five and forty-two years. The geneology of Huntington's disease sufferers in many parts of the world can be traced to emigrants from Europe in the 1600s. One particular lineage can be traced to brothers from Suffolk, England, who settled in Salem, Massachusetts, in 1630. It has been suggested that some "witches of Salem" may have suffered from this disease (Strange 1992, 187). Huntington's disease is characterized by progressive movement disturbances, dementia, and cognitive deficits.

Huntington's disease patients often have increased appetite and increased energy consumption, but they generally lose weight as the disease progresses. Their high rate of metabolism is attributed to constant, uncontrollable movements. However, low leptin levels (see chapter 14) have also been found, which indicate negative energy balance (Valenza and Cattaneo 2006). Greater body weight is associated with a slower rate of disease progression.

The cause of the disease is attributed to a mutation in a gene on chromosome 4 that codes for a protein given the name huntingtin. The normal gene has between 9 and 35 repeats of the DNA sequence CAG, which codes for glutamine in protein synthesis. The mutated gene generally has 40 to 60 CAG repeats, which produce a mutant form of the protein with dozens of glutamine residues. Transgenic strains of mice have been produced with the Huntington's disease gene, one strain with 115 CAG repeats and another with 150 CAG repeats.

The former animals show a slow progression of the disease, whereas the latter exhibit profound movement and cognitive anomalies at an early age (Maccarrone, Battista, and Centonze 2007).

It is also possible to produce experimental Huntington's disease in animals by exposing them to specific neurotoxins, such as 3-nitropropionic acid. This toxin impairs the mitochondrial respiratory chain by inhibiting the Krebs cycle enzyme, succinate dehydrogenase. This triggers a sequence of events that increase free-radical generation. The free radicals, in turn, generate signals for apoptosis, particularly in the striatum. Just how this neurotoxin produces the same effect as mutations in the gene for huntingtin is not known.

Huntingtin (the protein encoded by the defective gene in Huntington's disease) binds to phospholipids in the plasma membrane. It is believed to function in setting up microdomains within the plasma membrane (lipid rafts) that are responsible for transport and/or signaling in the synapse. Huntingtin becomes attached to palmitate (a saturated fatty acid), which anchors huntingtin protein to the membrane. When it gets anchored in the membrane, it is converted to a functional or active protein (see below). The polyglutamine extension on the mutated protein seems to impair palmitate attachment. The attachment of palmitate to proteins is a common mechanism for movement of signaling molecules into the nucleus to influence gene expression.

Elena Cattaneo and coworkers in Milan, Italy, used microarray analysis to find that mutant huntingtin affected expression of about two hundred different genes. Many of the affected genes are for proteins involved in cholesterol biosynthesis, including HMG-CoA reductase (the enzyme targeted by cholesterol-lowering statin drugs). They found reduced levels of mRNA for certain key enzymes in cholesterol biosynthesis prior to the onset of disease symptoms in mice having the gene for rapid disease progression. They also found that these mice had reduced levels of cholesterol in the brain and reduced sterol regulatory element binding protein (SREBP), which is responsible for moving transcription factors for proteins into the nucleus. Because the brain relies heavily on endogenous synthesis of cholesterol, these disruptions in cholesterol synthesis could be playing a critical role in the manifestations of Huntington's disease (Valenza and Cattaneo 2006). Perhaps physicians prescribing statins for lowering cholesterol should be aware of this relationship to Huntington's disease and monitor patients for such movement disorders.

Mauro Maccarrone's group (Maccarrone, Battista, and Centonze 2007) in Rome has focused on the role of cannabinoids in Huntington's disease. They found a reduction in the number of CB1 cannabinoid receptors (those located primarily in the brain) in both transgenic strains of Huntington's disease mice. The relative number of CB1 receptors corresponds with onset of disease. Some cannabinoids found in marijuana have been found to ameliorate symptoms of Huntington's disease and to be neuroprotective in 3-nitropropionic acid–induced disease in animals. Interestingly, cannabidiol, a component of marijuana that has little affinity for CB1 receptors and does not produce the psychotropic effects of tetrahydrocannabinol (THC), has been shown to protect against neuronal damage in animal models of Huntington's disease, probably through its antioxidant properties (Sagredo et al. 2007). The observed relationships of both cholesterol and cannabinoids to Huntington's disease beg the question of whether the cholesterol level in nerve tissue has some effect on activity of cannabinoid receptors.

These are just two approaches that are being used to better understand the underlying causes of this rare genetic disease, both involving lipids. It should not be surprising that lipids are involved in the manifestations of this disease, although to date there has been little reported on dietary manipulations or supplements that might be beneficial. The role of genetics and the proteins involved have captured most of the attention.

Alzheimer's Disease

In the early twentieth century, the German neurologist Alois Alzheimer first described the symptoms and neuropathology of a severe form of dementia, which has come to be the most common form of senile dementia. The disease is characterized by loss of memory and concentration, impaired language skills and muscular coordination, confusion, disorientation in space and time, mood changes, and general intellectual impairment. As the disease progresses, the patient is unable to perform normal daily tasks and may fail to recognize close family members. The disease appears most commonly in the elderly, especially after age seventy, although a genetic (familial) form of the disease appears at earlier ages. There is a nearly linear increase in prevalence of the disease from ages seventy to eighty-five, and it afflicts nearly 50 percent of the population by age eighty-five.

There was little progress in understanding the cause of Alzheimer's disease from the time of Alzheimer's first publications until the 1970s. There was a steady increase in incidence of the disease with increased life expectancy, particularly in industrialized countries. During the 1970s neuroscientists identified acetylcholine as a specific neurotransmitter that was most severely diminished in Alzheimer's disease. Specific acetylcholine neuronal tracts completely disappeared in advanced stages of the disease.

Postmortem analysis of brain tissue found significantly higher levels of aluminum in the brain tissue of Alzheimer's patients, which led to widespread speculation that dietary sources of aluminum should be avoided, such as aluminum cookware and aluminum hydroxide antacid tablets. In retrospect, the aluminum deposits in brain tissues of Alzheimer's patients are likely due to the strong affinity of aluminum for phosphates. In view of the large amount of phosphate in the debris of dying brain cells in Alzheimer's patients, it would be natural for any circulating aluminum to be trapped in that debris. It is now generally believed that aluminum is not directly involved in the neurodegenera-tion but arrives after the fact. There is also no greater incidence of Alzheimer's disease in regions of the world that have high levels of aluminum in drinking water (Rondeau 2002).

By the mid-1980s the nature of protein deposits found in the cellular debris of Alzheimer's disease brains was identified, namely amyloid plaque and neuro-fibrillary tangles. The beta-amyloid protein that forms the extracellular plaque was characterized as an aggregate of fragments of amyloid precursor protein (APP). The intracellular neurofibrillary tangles arose from tau protein, which is a structural protein forming neurofilaments. By the 1990s mutations in specific genes were identified in families that were prone to early onset Alzheimer's disease. Now, a century after Alzheimer's elaborate descriptions of behavioral changes and observed neuropathological changes at the microscopic level, we are beginning to understand the biochemical and molecular aspects of this devastating disease. However, there is still a long way to go before any effective treatments will be forthcoming, although there have been some encouraging reports that leave hope for optimism.

Neuropathology of Alzheimer's Disease

Alzheimer's disease is characterized by gross loss of neurons, particularly in the area known as the nucleus basalis of Meynert, which provides the primary cholinergic input to the cerebral cortex from the limbic system. There are specific microscopic lesions known as neuritic plaques and neurofibrillary tangles, primarily in the hippocampus and cerebral cortex. The neuritic plaques are accumulations of beta-amyloid protein, which arise from cleavage of amyloid precursor protein (APP) by a specific class of protease enzymes known as secre-tases. The neurofibrillary tangles are paired helical neurofilaments, consisting primarily of hyperphosphorylated tau protein, which is associated with micro-tubules and the basic molecular architecture of the intracellular space. Studies of the biochemical events that lead to these hallmark features of Alzheimer's disease have focused on these major protein abnormalities, in addition to the loss of acetylcholine or cholinergic neurons.

Although these are some of the more obvious anatomical features of Alzheimer's disease, a multitude of molecular and biochemical features are associated with them. Genetic studies have identified several genes that are linked to familial Alzheimer's disease, which accounts for only about 5 percent of all Alzheimer's cases. These include the genes for APP, presenilin, apolipo-protein E (apoE), and neurofilament light chain protein, and genes for synaptin, synaptophysin, and synaptosynapsin, which are structural proteins in synapses. There are a number of biochemical and physiological features in Alzheimer's disease that produce a dizzying array of changes in cell functions, including exacerbated inflammation, alterations in membrane enzymes and transport proteins, changes in structural and cytoskeletal proteins, altered calcium homeostasis and mitochondrial function, and changes in lipids and their bioactive metabolic products (Varadarajan et al. 2000).

A current theory regarding the progression of Alzheimer's disease is that APP, an integral membrane protein prevalent in the synaptic region of neurons, can be cleaved by specific proteases known as secretases. When cleaved by alpha-secretase, the cleavage product does not aggregate. However, when APP is cleaved by beta- or gamma-secretase, the protein fragments tend to aggregate to form senile plaque. In addition, these toxic beta-amyloid fragments can generate free radicals that promote several events that exacerbate the symptoms and pathology of Alzheimer's disease. It is possible that free-radical damage, such as peroxidation of lipid membranes and alteration of other biochemical components of cells (transport and structural proteins, enzymes, and DNA) may be an underlying mechanism in the progression of Alzheimer's disease. Oxidation of lipid membranes can cause cells to leak, which can trigger an immune response and its associated inflammation. Oxidation of membrane lipids could account for the observed alterations in calcium homeostasis as well as decreased levels of DHA found in Alzheimer's tissues (Oteiza et al. 1997). However, as mentioned in earlier chapters, DHA is not as susceptible to lipid peroxidation as other PUFAs, and the lower levels of DHA in the brains of Alzheimer's patients may have preceded the disease.

Inflammation in Alzheimer's Disease

The extracellular protein deposits and products of lipid peroxidation that may be floating around in these regions of the brain can trigger an immune response involving inflammatory mediators. Many clinical reports have indicated beneficial effects of nonsteroidal anti-inflammatory drugs (NSAIDs) in Alzheimer's disease, although the specific pathways for these beneficial effects are not certain. Several reports have conjectured a role for cyclooxygenase-2 (COX-2), which is the inducible form of COX that is prevalent in inflammatory conditions. It was suggested that COX-2-specific inhibitors might offer greater therapeutic value than nonspecific NSAIDs for Alzheimer's disease. However, there is no substantiated proof for greater efficacy of COX-2 inhibitors for treating Alzheimer's disease, and judging from their current status, they will probably not be found to offer any real benefit. Moreover, COX-1 is elevated in Alzheimer's disease and is constantly produced by neurons and glia, so nonspecific NSAIDs may indeed be more beneficial than COX-2 inhibitors (Yermakova et al. 1999).

COX-2 levels are also higher than normal in Alzheimer's brains and correlate with beta-amyloid protein levels, meaning that this enzyme is produced more as the disease progresses (Kitamura et al. 1999). However, COX-2 levels increase in the brain in a variety of disorders, including ischemia and seizures (Commentary 2000). Increased levels of COX, particularly COX-2, are generally an indication of an inflammatory response. Inflammation and its associated biochemical mechanisms are overtly toxic to cells. It has been found that inflammation and the characteristic pathology of Alzheimer's disease are localized in the same regions of the brain, namely frontal cortex and hippocampus. Inflammatory indicators are minimal in regions with the least susceptibility to Alzheimer's pathology, such as cerebellum.

Epidemiologic studies have shown generally favorable effects of long-term use of NSAIDs for arthritis in suppressing or delaying the onset of Alzheimer's disease, although the results are not conclusive (McGeer, Schulzer, and McGeer 1996). One of the problems in designing studies for definitive proof is that Alzheimer's disease is a slowly progressing disease, developing over years or perhaps decades. Postmortem analysis of nondemented elderly NSAID users compared with age-matched controls who were not chronic NSAID users showed similar amounts of senile plaque and neurofibrillary tangles. However, NSAID users had a significant reduction in the amount of activated microglia surrounding the senile plaques (Mackenzie and Munoz 1998). Activated microglia are a sign of inflammation, and NSAIDs are likely suppressing the inflammation that exacerbates the symptoms of Alzheimer's disease. Animal models have produced similar observations.

Although COX-1 and COX-2 levels are increased in specific regions of the brain in Alzheimer's disease, particularly those regions where Alzheimer's pathology is most prevalent, the role that COX and its lipid metabolites play in the progression of the disease is not well understood. It is quite possible that aberrant forms of APP and its cleavage products (beta-amyloid) that form extracellular aggregates of neuritic plaque are triggering an immune response that generates elevated levels of COX enzymes and their lipid metabolites.

Interestingly, transgenic mice that overexpress COX-2 have greater susceptibility to excitotoxins (Kelley et al. 1999). It has been suggested that increased levels of COX-2 in ischemia may promote cell death through production of free radicals (Pasinetti 1998). It is possible that NSAID inhibition of COX can lead to decreased production of apoptosis factors by neurons and less loss of the neurons that disappear in Alzheimer's disease.

Another related factor that enters the picture regarding lipids and Alzheimer's disease is the role of DHA-derived mediators. Recent work in the laboratories of Nicolas Bazan at Louisiana State University School of Medicine and Charles Serhan at Harvard Medical School has shown the effects of neuroprotectin D1 (NPD1), a DHA-derived lipid mediator, on human cells in culture. NPD1 promotes cell survival by inducing the expression of genes that suppress apoptosis. NPD1 reduces toxicity of specific amyloid proteins and has a general anti-inflammatory effect (Lukiw et al. 2005). As mentioned earlier (chapter 9), aspirin not only inhibits the normal enzymatic activity of COX, which produces proinflammatory prostaglandins, but also switches the activity of COX-2 to catalyze production of docosatrienes from DHA, namely NPD1 (Serhan et al. 2004).

Since the brain has the most abundant deposits of DHA in the body, other than the retina of the eye, this effect of aspirin should be most pronounced in the brain. NPD1 was also shown to regulate secretion of beta-amyloid peptides (precursors of senile plaque), which in turn decreases the amount of inflammation and promotes cell survival. All these effects would diminish the loss of neurologic functions caused by Alzheimer's disease. Lower levels of DHA in Alzheimer's brains would result in less metabolism of DHA by these pathways and hence less protection.

Some studies have shown that lipoxygenase (LOX) products, particularly 12-HETE (12-hydroxyeicosatetraenoic acid; see chapter 3 and appendix C), promote apoptosis in beta-amyloid-exposed rat neurons grown in culture (Lebeau et al. 2004). This would indicate that, in addition to COX inhibition, lipoxygenase inhibition may be necessary to prevent overt loss of neurons and neurologic function in Alzheimer's disease. Both of these enzymes (COX and LOX) catalyze the conversion of LCPUFAs to bioactive eicosanoids. An Asian herbal medicine (Sho-Saiko-To) containing the active ingredient baicalein, which is used for the treatment of chronic hepatitis (inflammatory liver disease), was found to inhibit 12-lipoxygenase (Shimizu et al. 1999). There have been suggestions that this herb could be beneficial in slowing the progression of Alzheimer's disease, although studies to that effect have not yet been published.

A large, long-term (twenty-seven year) prospective study showed some correlation between middle-age obesity and the risk of dementia later in life (Whitmer et al. 2005). Obese people (BMI greater than 30) had a 74 percent greater risk of developing dementia, and overweight people (BMI = 25 to 29.9) had a 35 percent greater risk of developing dementia compared with normal-weight individuals (BMI = 18.6 to 24.9). There are many biochemical factors associated with obesity that could play a role in this observed relationship. These include, but are not limited to, increased inflammation and proinflammatory signaling factors in obesity. This is supported by other observations that indicate metabolic syndrome (associated with obesity) increases risk of cognitive decline (van den Berg et al. 2008).

When a transgenic strain of mice that overproduces amyloid precursor protein (APP) was fed a high-fat diet to induce insulin resistance (another syndrome associated with obesity in humans), it was found that gamma-secretase, the enzyme that cleaves APP, became more active in producing the toxic form of beta-amyloid protein. In addition, there was a decrease in insulin-degrading enzyme, which is normally found in the brain and is activated when insulin binds to its receptor. The insulin-degrading enzyme has also been found to degrade beta-amyloid protein, thus decreasing the toxic effects and ultimate accumulation of this protein as senile plaque. These are very significant findings in view of the fact that insulin resistance, or type 2 diabetes, is associated with a two- to threefold increased risk of developing

Alzheimer's disease (Ho et al. 2004). Although this may be the connection that relates obesity to increased risk of later dementia, there are many other possibilities. The role of dietary sugars, particularly fructose, may be another factor that needs to be investigated, since high-fructose diets are known to increase the incidence of type 2 diabetes (see chapter 14).

Effect of Statins in Alzheimer's Disease

Cholesterol-lowering drugs (statins) can decrease beta-amyloid production in vitro and in animal models of Alzheimer's disease (Kaether and Haass 2004). This is probably due to an increase in the cleavage of amyloid precursor protein (APP) by alpha-secretase, the enzyme that produces a nontoxic or nonamyloid protein product (Kojro et al. 2001). Inhibition of the enzyme that converts cholesterol to cholesteryl esters (acyl-CoA:cholesterol acyl transferase, ACAT) also decreases the amount of beta-amyloid deposition in the mouse model for Alzheimer's disease (Hutter-Paier et al. 2004). Some epidemiological studies and clinical trials have indicated that statin use can lower the risk of Alzheimer's disease (Wolozin et al. 2000). Since statins reduce inflammation (chapter 8), it is not surprising that these lipid-lowering drugs could curtail the pathology of Alzheimer's disease. However, it is important to be aware of the side effects of these drugs, including cognitive impairment, which could have detrimental effects on the patient's overall quality of life in the early stages of the disease.

The active forms of the deleterious gamma-secretase (composed of presenilin) and beta-secretase (or BACE) are located in lipid rafts within membranes of cells (Vetrivel et al. 2004). It is thought that if APP comes in contact with lipid rafts containing this pair of secretase enzymes, they will cleave the protein into its toxic beta-amyloid products. This would promote amyloid accumulation and subsequent neuronal death and cognitive impairment. It is not clear what roles the secretase enzymes play in the normal scheme of brain function, but it seems unlikely that they are there to cause trouble by their actions in producing the toxic beta-amyloid proteins. It is also not clear whether lipid rafts, where the detrimental secretase enzymes are located, are playing a role in the etiology of Alzheimer's disease.

Certain genetic variations of apolipoprotein E (apoE), which is a component of very low-density lipoprotein (VLDL) as well as low-density lipoprotein (LDL), have been found to correlate with Alzheimer's disease. In particular, there is a much higher frequency of the apoE4 genetic variant in people with Alzheimer's disease than of the other two main variations (apoE2 and apoE3) (Strittmatter and Roses 1995). There is evidence that apoE2 and apoE3, but not apoE4, bind to tau protein, which is a component of neurofibrillary tangles in Alzheimer's disease. It is postulated that apoE2 and apoE3 block aberrant interactions of neurofilaments, preventing them from developing into neurofibrillary tangles. However, other interactions of these different forms of apoE proteins could be contributing to the pathogenesis of Alzheimer's disease.

In conclusion, there is good evidence that eicosanoids formed from polyunsaturated fatty acids are involved in the progression of Alzheimer's disease and that COX inhibitors or NSAIDs may slow this progression. Inflammation seems to play a part, and dietary constituents that suppress inflammation would probably be beneficial in slowing the advances of many degenerative diseases. Conversely, the omega-6 oils, which fuel the inflammatory response, could augment these diseases. Although statins have shown positive effects in slowing the progress of some neurodegenerative diseases, their favorable effects are probably via mechanisms other than lowering cholesterol, such as anti-inflammatory effects.

Any beneficial effects of diet are likely to be gained prior to the onset of symptoms. There might be little to be done through dietary practices once the symptoms have begun to appear, primarily because most of the damage to tissue has already been done. There is not much chance for repair in the central nervous system, particularly in old age. However, moderate increases in omega-3 oils, along with decreases in omega-6 oils, would have few undesirable side effects and may slow the progression of many degenerative diseases, particularly where inflammation is involved.

Parkinson's Disease

In the early nineteenth century, James Parkinson gave detailed descriptions of the disease he called shaking palsy, which came to bear his name. The onset for Parkinson's disease is generally after sixty years of age, and it afflicts about 0.15 percent of the population, with little or no difference among various ethnic groups. Less than 10 percent of cases appear to be genetic in origin. Inherited Parkinson's disease tends to have an earlier age of onset than the idiopathic form. The classic symptoms of Parkinson's disease are tremor (while resting or idle but not during movement or sleep), rigidity (resistance to passive movement), and disturbances in movement (slow movement or bradykinesia). Other symptoms include stooped posture, speech problems, and small handwriting (micrographia). There is also increased risk of dementia.

The neuropathological changes include gross degeneration (loss of cells) of the substantia nigra and appearance of abundant deposits of cell debris known as Lewy bodies in the substantia nigra and locus ceruleus. Lewy bodies are not restricted to these brain regions. The protein alpha-synuclein is a major component of the cell debris in Lewy bodies. This protein is the product of one of the genes associated with hereditary Parkinson's disease. Aggregation of alpha-synuclein stimulates transformation of microglia to macrophage-type cells. The resulting generation of reactive oxygen species leads to inflammation and further damage in a localized region of the brain. The orchestration of these events is analogous to that observed in Alzheimer's disease, albeit with a different protein (beta-amyloid in the case of Alzheimer's disease), a different neurotransmitter system being affected (acetylcholine in Alzheimer's), and different loss of function resulting from the ensuing inflammation and degeneration of cells.

There is profound depletion of dopamine in the caudate nucleus, putamen, and substantia nigra, and to a lesser extent in the globus pallidus. About 80 percent of the dopamine in striatum (caudate nucleus and putamen) disappears before symptoms of the disease are noticed. Positron emission tomography can be used to measure dopamine loss in Parkinson's patients. One of the early treatments for Parkinson's disease was L-DOPA (L-dihydroxyphenylalanine) therapy. L-DOPA, which is an amino acid similar to tyrosine, can cross the blood-brain barrier via the large neutral amino acid transporter and is metabolized to dopamine. In fact, L-DOPA is a normal intermediate in the metabolism of tyrosine to dopamine in the brain.

It is not known what causes the loss of dopamine and dopamine neurons in these brain regions (most brain dopamine is in these regions), but there is much speculation that environmental factors may contribe to this loss. An environmental cause is supported by the discovery around 1980 that a substance called MPTP (N-methyl-4-phenyltetrahydropyridine), found as a by-product in a synthetic street drug in the San Francisco Bay area, produced all the classic symptoms and pathology of Parkinson's disease in a population of young adults using the illicit drug (Langston et al. 1983). It was eventually found that MPTP is metabolized by monoamine oxidase B to the active neurotoxin MPP+ (N-methyl-4-phenylpyridinium ion), which is then taken into dopamine neurons by the dopamine transporter (hence the selectivity for dopamine neurons). In the dopamine neurons MPP+ inhibits the mitochondrial respiratory chain, which leads to the depletion of energy stores. It also generates free radicals in the process, causing destruction of the cells.

MPP+ is structurally similar to the herbicide paraquat, which was used in no-till farming in the United States in the 1970s. Paraquat was also used to eradicate marijuana crops in Mexico. Unfortunately, Mexican farmers salvaged their wilted crops and sold them to dealers in the United States, whence the paraquat came, causing serious lung damage to anyone smoking the tainted marijuana. However, the ionic nature of paraquat precludes it crossing the blood-brain barrier under normal conditions, so the relevance of its ability to destroy neurons when injected directly into the brain has been questioned (Miller 2007).

A Parkinson-like syndrome or parkinsonism was also observed in manganese miners in Chile in the 1960s (Cotzias 1971). These miners were treated with L-DOPA and found to respond as well as Parkinson's disease patients. Exposure to manganese dioxide dust induces the parkinsonian syndrome and degeneration of dopamine cells in the striatum, but divalent manganese ion dissolved in water does not.

There was also an ALS-parkinsonian syndrome reported on the island of Guam. This unusual syndrome was traced to the presence of a neurotoxin, beta-N-methylaminoalanine, in seeds from the cycad bush that were ground into flour and used for food during times of food scarcity on the island (Murch et al. 2004). These are just a few examples of environmental factors that can produce Parkinson-like symptoms.

There is much evidence that Parkinson's disease is caused by oxidative stress resulting in degradation of dopamine neurons, with lipid peroxidation of membranes being one eventuality. Inflammation is invoked as the source of free radicals in the idiopathic form of the disease, which may entail production of nitric oxide and a range of reactive oxygen species (Knott, Stern, and Wilkin 2000). There is also speculation that glutamate-related excitotoxicity may play a role. Genetic studies indicate that alpha-synuclein may be abnormal, at least in one genetic form of the disease. Misfolding and aggregation of this protein could lead to attack by microglia, resulting in inflammation (W. Zhang et al. 2005). Although radical-mediated damage has been invoked, studies designed to diminish oxygen radical production by inhibiting monoamine oxidase B with selegiline (deprenyl) along with vitamin E as an antioxidant have not been convincing. Some argue that levels of vitamin E supplements used may have been too low to be effective. Vitamin E was found to slow the progression of Alzheimer's disease, but at much higher doses (Mandel et al. 2003).

The involvement of inflammation in both diseases suggests similar roles for lipid mediators of inflammation (prostaglandins, leukotrienes, and lipoxins). The use of NSAIDs to suppress inflammation and slow the progression of disease would likely apply to Parkinson's disease as well. There seem to have been fewer studies of NSAIDs and omega-3 oils in relation to Parkinson's disease compared with Alzheimer's disease, although the similarities in etiology would recommend similar strategies.

Cannabinoids are another group of lipid messengers that may have a detrimental influence on Parkinson's disease. As mentioned earlier, these retrograde lipid messengers generally travel from the postsynaptic neuron back to the presynaptic neuron, where they decrease firing of the latter. This action is similar to the effect of dopamine in the synapse, which can bind to presynaptic (D2) dopamine receptors and cause a decrease in further dopamine release. In the case of Parkinson's disease, in which dopamine neurons have disappeared, the aim is to get the remaining neurons to release more dopamine. Consequently cannabinoids, whether from endogenous sources or as a result of consuming marijuana or pharmaceutical Marinol, would likely exacerbate the symptoms of Parkinson's disease. There do not appear to be any published reports of such effects of cannabis on Parkinson's disease.

Conclusions

Inflammation is inherently present in all the neurodegenerative diseases discussed above, and it seems that anti-inflammatory agents could help to slow the progression of most of them. There is some evidence that frequent use of anti-inflammatory agents (NSAIDs) for arthritis and other inflammatory syndromes may slow the progression of Alzheimer's disease, although placebo-controlled studies of these drugs may not be practical in view of the slow progression of Alzheimer's. Animal models for both amyotrophic lateral sclerosis (ALS) and Alzheimer's disease indicate a protective effect by agents that suppress inflammation. The suppression of inflammation by omega-3 oils suggests that a more favorable ratio of omega-3 to omega-6 oils in the diet throughout life could help to delay the onset or suppress the progression of these devastating neurodegenerative diseases.

PART THREE

Influence of Diet
on Overall Health

14

Obesity

Health Consequences and Dietary Influences

The incidence of obesity has escalated dramatically in the United States in the past twenty years to reach epidemic proportions. This obesity epidemic defines a generation with increased risks of incurring a wide range of unhealthy effects, including high blood pressure, type 2 diabetes, coronary heart disease, stroke, gall bladder disease, and several cancers, especially of the breast, prostate gland, and colon. Obesity has become one of the leading causes of preventable early death in the United States. This chapter will look at what the possible underlying causes are. It will be helpful first to define what is meant by obesity and then to look at how its occurrence has increased in recent years.

Obesity is defined as having an abnormally high proportion of body fat and is predicated on body mass index (BMI), which is calculated using a simple formula based on height and weight:

BMI = weight (in kilograms)/height (in meters) squared; or BMI = kg/m^2.

BMI can be calculated using pounds for weight and inches for height by using the conversion factor 704.5:

BMI = 704.5 × weight (in pounds)/height (in inches)^2.

Some organizations, such as the American Dietetic Association, use 700 for the conversion factor, since BMI provides only a rough assessment of ideal body weight relative to overweight or obese body weight. The National Institutes of Health (NIH) adopted definitions for overweight and obesity in adults similar to those used by the World Health Organization (WHO): BMI between 25 and 30 is classified overweight, and BMI greater than 30 indicates obesity. There is also increasing concern regarding childhood obesity, although the federal government does not have a generally accepted definition for obesity in children and adolescents in terms of BMI.

National statistics for the United States indicate that the prevalence of overweight adults in the population changed very little from 1960 to 2000 (from 31.5 to 33.6 percent), whereas obesity more than doubled during that time period and now afflicts more than 30 percent of the adult population (WIN-NIDDK 2007). Extreme obesity, defined as a BMI greater than 40, increased nearly sixfold between 1960 and 2000 (from 0.8 to 4.7 percent of the adult population). The largest rate of increase occurred during the last decade of the twentieth century.

It is generally agreed that obesity increases one's risk of health problems. What this really means is that a greater proportion of obese people have health problems relative to the population whose body weight falls within the ideal range (BMI between 18 and 25). A few examples will serve to illustrate. Two-thirds of the seventeen million people in the United States with type 2 diabetes have a BMI greater than 27, and nearly 50 percent have a BMI greater than 30 (WIN-NIDDK 2007). The prevalence of hypertension (systolic blood pressure greater than 140 and diastolic pressure greater than 90 mm Hg) is about 42 percent in obese men and about 38 percent in obese women. There is a strong correlation between blood pressure and BMI; as BMI increases, blood pressure tends to increase (see figure 14.1).

Nearly half of postmenopausal women with breast cancer have a BMI greater than 29. Having a high BMI shortens one's life expectancy by several years, with

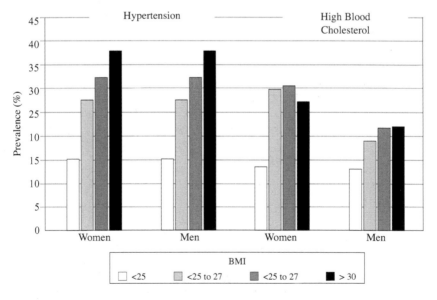

FIGURE 14.1. Relationship of body mass index (BMI) to hypertension (cystolic pressure > 140 and diastolic pressure > 90) and high blood cholesterol (> 240 mg/dL).

Data from WIN-NIDDK 2007.

most of the increased risk of death at an earlier age coming from cardiovascular causes. It is interesting to note that high serum cholesterol levels are associated with being overweight and obese, but the prevalence of high cholesterol does not increase with BMI, as some other risk factors do (compare blood pressure, figure 14.1). The prevalence of high total serum cholesterol (greater than 240 mg/ dL) among obese individuals is 22 percent for men and 27 percent for women, and varies little for high overweight and low overweight categories (figure 14.1). In contrast, the prevalence of high cholesterol for those who are not overweight (BMI less than 25) is only about 13 percent for men and women.

Changes in Eating Patterns

Some view the obesity epidemic as a food fight, as the title of one book suggests (Brownell and Horgen 2004). On one side of the issue are the advocates for healthier foods in general, but especially in schools. They argue that the food industry uses slick and deceptive marketing to sell the public, especially children and teenagers, snacks and fast foods that are high in calories from fats and sugars and low in vitamins, minerals, and other essential nutrients. On the other side of the table is the food industry, claiming that it is unfairly targeted, because it simply offers the public what they want; the purveyors can't help it if the lazy populace won't get off their obese derrières and exercise to burn off the excess calories they are eating.

Many argue that the obesity epidemic is a class issue (Wang 2001). It seems the lower economic classes are the most prone to obesity in America, which seems to defy logic. How can those who have the least money consume the most food calories? In this land of plenty there were about thirty-six million people living in poverty in 2004, according to the U.S. Census Bureau, including about thirteen million children, and millions more struggling to stay above the official poverty line (U.S. Census Bureau 2005). The lower classes tend to buy foods that are the easiest to prepare for what is perceived to be the lowest price. This usually means packaged, prepared foods that require little or no preparation and little or no cooking, which often tend also to be the least nutritious.

Heads of households, who are responsible for providing meals for the family after working extended hours in the day to make ends meet, frequently resort to fast foods or national restaurant chains that pride themselves on the largest portions for the lowest prices. The documentary film *Supersize Me*, by Morgan Spurlock (2004), exposed the gluttonous horrors associated with one chain's marketing gimmick of supersized portions. It is human nature to want to get the most for your money, and if you cannot afford to buy quality, the next best thing in many consumers' eyes is to purchase quantity.

Several studies have reported an increase in the relative size of food portions over the past quarter century (Young and Nestle 2002). It seems market researchers and marketers have convinced providers that this is what the public wants. Providers compete with one another to offer the biggest portions. Children's portions are sometimes available but are often not publicized. Adults are usually discouraged from getting children's portions and often denied them. Barbara Rolls and colleagues at Pennsylvania State University found that subjects consumed significantly more calories when larger portion sizes were offered (Rolls, Morris, and Roe 2002).

Another consideration is food content. Many people focus on calories in fats, since each gram of fat yields about 9 kilocalories (kcal), whereas proteins and carbohydrates contain only about 4 kcal per gram. Most health and nutrition professionals condemn the average American for consuming about 40 percent of calories as fats, recommending that fats constitute no more than 30 percent of calories in the diet. Yet 40 percent of calories derived from fats has been the norm in America for the past forty years at least, certainly well before the sharp rise in incidence of obesity took place. So it seems unlikely that percentage of fats in the diet is the cause of obesity. The argument that this high proportion of fats in the diet is unhealthy and increases risk for several diseases is not well grounded in scientific studies. There will be more discussion of high-fat diets in the next chapter.

The proportions of the other two major macronutrients, proteins and carbohydrates, have not changed much either. Protein consumption in the United States is higher than in many other countries because of the amount of meat in the diet and has remained fairly constant at about 15 percent of total calories for the past half century. The remainder, about 45 percent of calories, comes from carbohydrates. Although the percentage of carbohydrates in the diet has not changed much, the composition of carbohydrates has changed significantly and is likely a major factor contributing to obesity. Although obesity seems to be caused by the recent increase in total food consumption as a result of larger portions and increased frequency of snacking, the role of carbohydrates and fats in triggering hunger and satiety cannot be ignored (see below).

The Changing Face of Dietary Carbohydrates

Sugar, extracted from sugarcane, was a rare and expensive commodity in Europe before Columbus discovered America, with ten pounds of sugar priced at more than 25 percent of the price of an ounce of gold. Sugar was perhaps the most instrumental crop in expanding the African slave trade, especially to the Caribbean basin (Hobhouse 1985). When Europeans discovered that sugar could be extracted from sugar beets in the late eighteenth century, the price of sugar plummeted. Today the price of sugar is a fraction of what it was in 1400, relative to the price of an ounce of gold.

A century ago the proportion of complex carbohydrates to simple sugars in the American diet was much higher than it is today. In addition there was a dramatic shift in processed sugar consumption in the late 1970s. First there was a move in the beverage industry to introduce diet sodas with artificial sweeteners for calorie-conscious consumers. Diet drinks accounted for about one-fourth of soft drinks sold in the United States in 1997 (Putnam and Allshouse 1999). However, total consumption of soft drinks went up dramatically in the 1980s and 1990s, and most of the sweetener added to nondiet drinks was high-fructose corn syrup.

High-fructose corn syrup became a cheaper sweetener than sucrose in the 1970s, owing to availability of waste products from the corn oil industry. Sucrose, or table sugar, is extracted from sugarcane (a tropical grass) or sugar beets (a root vegetable grown in temperate climates). One does not think of getting sugar from corn (maize), even though hybrid varieties have been developed for the "sweet corn" that is sold by the ear during summer months in American markets and rural roadside stands. Green plants generally make sucrose as a storage sugar, but the levels of sucrose in most plants, including corn, does not make it worth extracting. High-fructose corn syrup is a technological feat of the food industry, not a quirk of maize genetics.

The American food industry extracts large quantities of oil from corn and other grains, which is apparent when looking at the vegetable oil section of any major supermarket. Corn oil is generally hydrogenated to produce shortening and margarine, and much of this partially hydrogenated vegetable oil is used in processed foods. After the corn kernels are pressed to extract the oils, starches, fiber, and protein are left in the mash. The carbohydrate to protein ratio in corn is more than six to one, whereas in kidney beans or soybeans the ratio is nearer three to one. This means that the protein content of corn is very low, about one-half that of beans. The corn mash is predominantly starch, which for years was sold as cornstarch. There was not a big demand for cornstarch, and the corn mash was of too poor quality to be used as animal food because of its low protein content. During the nineteenth and twentieth centuries, Americans upgraded the worth of their corn mash by fermenting it and distilling the culture to make corn whiskey or bourbon, the latter of which is widely considered to be the hallmark of American liquors.

For more than a halfcentury cornstarch was treated with enzymes to convert it into sugar, primarily glucose and maltose, and sold as corn syrup. However, it was discovered that treatment with another enzyme converts most of the glucose into fructose. Perhaps the original idea was to get yields of fructose and glucose in proportions that would closely resemble those in the sucrose molecule and some other natural sweeteners, such as honey, which has about equal proportions of fructose and glucose. Fructose happens to be about 70 percent sweeter than sucrose, nearly 2.5 times sweeter than glucose, and about 5 times sweeter than maltose.

High-fructose corn syrup ranges from about 55 percent to as high as 80 percent fructose, with the remainder being mostly glucose and minor amounts of maltose and other sugars (Sanda 2004). It made sense to convert the hydrolyzed cornstarch to larger proportions of fructose if the aim was to get a sweeter product. By using a sweeter sugar (fructose), it would be possible for manufacturers of processed foods to use less sweetener (in terms of quantity and calories) in their products to achieve the same sweetness at lower cost. Yet levels of sweetness did not stay the same. A standard 8-ounce serving of cola in the 1960s contained between 85 and 105 kcal from sucrose, while soft drinks today contain between 112 and 135 kcal per 8-ounce serving from high-fructose corn syrup, in addition to being sold in 16- to 20-ounce containers that contain up to 330 kcal per 20 ounces.

The use of sweeteners increased by more than 22 percent between 1985 and 2005, with high-fructose corn syrup accounting for all this growth (Haley et al. 2005). In fact, refined sugar (sucrose) usage showed a modest decline during that period (see table 14.1). Perhaps an even more telling picture is given by consumption of carbonated soft drinks, which increased 51 percent between 1980 and 1998 (U.S. Census Bureau 2000, 148). Sales of diet sodas more than

TABLE 14.1.

Statistics on per capita consumption of sweeteners, fat, and protein

Commodity	1980	1998	% change
Refined sweeteners	55.8 kg	69.9 kg	30
Sugar	38.0 kg	30.4 kg	−20
High-fructose corn syrup	8.6 kg	29.0 kg	336
Carbonated soft drinks (total)	133 L	201 L	51
Regular sweetened	113 L	157 L	38
Diet	19.3 L	43.9 L	120
Uncarbonated soft drinks	24.2 L (1990)	34.4 L (1997)	42
Total carbohydrate consumption	395 g	509 g	29
Total protein consumption	96 g	112 g	17
Total fat consumption	149 g	156 g	5
Total calories	3,300 kcal	3,800 kcal	15

Source: U.S. Census Bureau 2000, 146.

doubled during that time period, while those of sugar-sweetened carbonated drinks rose by 38 percent (table 14.1). During the 1990s uncarbonated soft drinks, such as fruit punches, iced tea, and lemonade, were introduced on a large scale. Many of these drinks are also sweetened with high-fructose corn syrup, in addition to natural sugars in juices added to some of them.

On average, the daily carbohydrate consumption of each American increased 29 percent between 1980 and 1997, while protein consumption increased 17 percent and fat consumption increased 5 percent. Total calories consumed per day increased from 3,300 to 3,800 kcal (15 percent) during this time period. Annual refined sugar (cane and beet) consumption decreased by 16.6 pounds per capita between 1980 and 1998, while that of high-fructose corn sweeteners increased by 43.5 pounds (U.S. Census Bureau 2000, 147). Energy expenditure did not keep pace with increased caloric intake, so the excess energy was stored in the most efficient form possible—as adipose fat.

Metabolic Consequences of Fructose

The statistics alone indicate that the greatest contribution to increased calorie consumption is coming from carbohydrates, with the largest amount from high-fructose corn syrup. There have been numerous scientific studies to determine the health effects of fructose in the diet, relative to other sugars. Nearly a century ago Harold Higgins showed that fructose was converted to fat more easily than glucose (Higgins 1916). Work by Ian Macdonald in London in the mid-1960s compared the effects of different dietary carbohydrates on serum lipids in human subjects. The original aim was to determine whether sucrose has a more detrimental effect on serum lipids than starches. Since sucrose causes an increase in serum lipids, whereas starch, maltose (an enzymatic breakdown product of starch), and glucose (the ultimate breakdown product of starch) do not, it became imperative to determine whether fructose was causing this effect. When sucrose is digested it produces equal portions of glucose and fructose. Sure enough, fructose was found to increase levels of triglycerides in the blood of men and postmenopausal women, but it decreased triglycerides in younger women (Macdonald 1966). These studies were conducted before separation of different lipoprotein fractions (HDL, LDL, and VLDL) became routine.

Studies done in the early 1980s at Beltsville Human Nutrition Research Center of the U.S. Department of Agriculture confirmed that fructose consumption (single dose) resulted in higher serum triglyceride levels in men. When subjects were divided into hyperinsulinemic and normal insulin groups, a dramatically elevated triglyceride level (60 percent increase) was observed in hyperinsulinemic individuals while there was no significant change in normal insulin subjects. In 1983 the authors concluded that moderate amounts of fructose in the American diet can produce undesirable changes in blood lipids that are associated with heart disease and that hyperinsulinemic men are more susceptible to these changes than men with normal insulin levels. The hyperinsulinemic condition is also known as insulin resistance and is often a prelude to type 2 diabetes. Clearly fructose, whether from sucrose or high-fructose corn syrup, cannot be ignored as a dietary factor that increases risk of developing heart disease by affecting serum lipids. Fructose is probably doing much more than merely raising serum triglycerides to aggravate conditions that promote heart disease.

The increase in blood triglycerides with high-fructose diets is readily explained by fundamental biochemical principles. A transporter protein (GLUT5) that does not depend on insulin for activity brings fructose into cells. When fructose enters the liver after being absorbed into the bloodstream, it will be converted preferentially to fructose-1-phosphate and metabolized primarily to pyruvate through the glycolytic pathway—the pathway for carbohydrate metabolism to produce energy under anaerobic or aerobic conditions. Glucose is preferentially converted to glucose-6-phosphate when it enters cells, which can be stored as glycogen or enter several other metabolic pathways that are essential for cell survival. Without going into details of these biochemical pathways, suffice it to say that fructose is preferentially metabolized to fats.

On the other hand, glucose is preferentially metabolized to glycogen until the glycogen stores are filled, and then any excess will be converted to fats. Consequently, when someone is consuming a meal with some starch and washes it down with a high-fructose soft drink, the fructose will tend to be converted immediately to fats. Glucose from the soft drink sweetener and from breakdown of starch will be stored as glycogen and then converted to fats only when there is excess.

There is a genetic disorder known as fructose intolerance that can result in severe hypoglycemia as a result of consuming excessive fructose (Devlin 2002, 616). Prolonged ingestion of fructose by young children who have this condition (although their parents may be unaware of it) can lead to serious health and even life-threatening consequences. Liver cells become depleted of inorganic phosphate owing to formation of large amounts of fructose phosphate. The depleted inorganic phosphate status can suppress the glucagon-induced response of liver cells to secrete glucose when blood glucose levels are low. Release of glucose from glycogen (glycogenolysis) requires inorganic phosphate. Low inorganic phosphate also curtails formation of ATP from ADP in normal cellular respiration. Even in normal individuals, high levels of fructose could reduce levels of ATP and inorganic phosphate, which would likely alter normal metabolism of glucose and intracellular controls involved in glucose distribution and homeostasis.

There is evidence that dietary fructose promotes insulin resistance and type 2 diabetes (Basciano, Federico, and Adeli 2005). Type 2 diabetes was originally called adult-onset diabetes, but this latter label is being used less frequently because it is becoming more prevalent in children and adolescents. Normally, when glucose levels increase in the blood, beta cells in the pancreas secrete insulin, which in turn triggers muscle cells and adipose tissue to remove more glucose from the blood and bring the glucose level back to normal. This is done by activation of a glucose transporter (GLUT4) in these cells, which is a protein that carries glucose across cell membranes. The transporter does not work until insulin binds to its receptor. The GLUT4 transporter can be taken out of the plasma membrane of cells when blood glucose levels are low by internalization of membrane vesicles containing the transporter. Phosphorylation of proteins in response to insulin results in the transporter being returned to the plasma membrane (Watson and Pessin 2001).

A study from Japan (Hyakukoku et al. 2003) showed that there was diminished phosphorylation (attachment of phosphate) to certain proteins involved in the insulin response, particularly in muscle tissue, when rats were fed a high-fructose diet. Although this indicates that fructose can interfere by this mechanism, there are certainly other ways that fructose can diminish the insulin response.

Fructose can initiate free-radical reactions much more than glucose. Fructose is more active than glucose at cross-linking and polymerization of proteins as well as modification of proteins by forming advanced glycosylation (or glycation) end products (AGEs) (Sakai, Oimomi, and Kasuga 2002). Such reactions can result in modification of proteins, including enzymes, transporters, structural proteins, and other types. Fructose inactivates antioxidant enzymes, such as superoxide dismutase and catalase, among others (Zhao, Devamanoharan, and Varma 2000). It is not clear how it inactivates these enzymes, but it is probably due to formation of reactive fructose breakdown products that bind to them and inhibit them.

Formation of glycated proteins is common in diabetics and is the cause of many physiological complications of diabetes, such as cataracts and destruction of hemoglobin. Glycation of lens proteins in the eye causes cataracts and can result in loss of vision if it is too extensive. Glycation of hemoglobin diminishes oxygen-carrying capacity and results in fatigue from oxygen depletion in tissues. Interference with oxygen binding to hemoglobin results in more free oxygen floating around in the blood to wreak havoc by causing unwanted free-radical reactions (as described in chapter 5). Other proteins in various tissues exposed to fructose and its oxidation products could also become glycated. If the glycated proteins are receptors for insulin or other hormones, glycation will likely result in a diminished response to those hormones.

Hyperglycemia is known to increase the formation of oxidized LDL as well as glycated LDL, which are both scavenged by the macrophages that cause atherosclerosis (Veiraiah 2005). In view of the fact that fructose is much more active at forming AGEs than glucose, increased fructose consumption probably produces more of the oxidized and glycated LDL, leading to more severe atherosclerosis. AGEs will also increase damage to vascular endothelial cells, making the lining of arteries much less resilient and more fragile. Modification of proteins in the vascular lining could also incite the immune system to attack and promote inflammation that can lead to a heart attack.

Degradation of fructose results in formation of significant amounts of formaldehyde (Lawrence, Mavi, and Meral 2008), which is a highly reactive compound and classified as a carcinogen. Formaldehyde can cause respiratory problems, including asthma. It was previously thought that the most common route of exposure to formaldehyde was by inhalation, but the recent finding that it can be present in soft drinks may change this viewpoint. Is it just a coincidence that asthma cases have skyrocketed after high-fructose corn syrup was introduced and became the most common sweetener in soft drinks? Unfortunately, there is not an official limit regarding how much formaldehyde can be in soft drinks. It is surprising that there was so much concern about removing formaldehyde from indoor air several years ago (as well as from academic laboratories for biology after 2001), yet there does not seem to be much interest in getting it out of soft drinks.

The fact that fructose can initiate free-radical reactions means that it can also cause damage to cell membranes by lipid peroxidation (Lawrence, Mavi, and Meral 2008). It can also cause or promote genetic mutations, which would lead to increased risk of cancer. Many studies indicate that fructose is doing much more widespread damage in the body than glucose and will promote a wide range of adverse health effects, such as premature aging, cardiovascular disease, metabolic disorders, and cancer (Stanhope and Havel 2008).

Fructose also increases plasma levels of uric acid, which can lead to gout (Macdonald, Keyser, and Pacy 1978). It is not a coincidence that fructose can promote the same disorders that are seen in metabolic syndrome or syndrome X, a condition that is generally associated with obesity and has risen sharply since the introduction of high-fructose sweeteners (Elliott et al. 2002). Is high fructose the cause of these diseases? Some may argue that it has not been proven, but the evidence strongly supports a role for fructose in a wide range of disorders. See table 14.2 for important metabolic consequences of fructose compared to glucose.

Metabolic Syndrome

As a result of dietary fructose increasing fat production in the liver, there is a concomitant increase in plasma trigylcerides due to liver secretion of very low-density lipoprotein (VLDL). VLDL is a serum lipoprotein that carries mostly triglycerides and relatively little cholesterol compared with LDL. Since sweeteners such as sucrose and high-fructose corn syrup also contain glucose, there is a rise in blood glucose that parallels the increased fructose. The rise in glucose triggers release of insulin from the beta cells in the pancreas and facilitates glucose uptake from the blood into many different tissues. When there

TABLE 14.2.

Important metabolic consequences of fructose versus glucose

Parameter	Fructose	Glucose
Primary metabolic product	Fat	Glycogen (starch)
Serum triglycerides (VLDL)	Increase	No change
Acute serum insulin levels	No change	Increase
Long-term insulin levels	Hyperinsulinemia	No change
Long-term glycemic effect	Hyperglycemia	No change
Liver inorganic phosphate level	Decrease	No change
Potential for glycation of proteins	High	Low
Free-radical generation	Highest of any sugar	Lowest of any sugar[a]
Production of formaldehyde (in drinks)	Measurable	Not measurable

Source: Compiled by the author from information in the sources cited in the text.

[a] Glucose was lowest of any monosaccharide tested, but sucrose also produced low levels.

are high levels of triglycerides (or VLDL) in the blood as a result of elevated fructose consumption, muscle cells will utilize this source of energy when it is available. If a muscle is not active, it will take a long time for it to metabolize the absorbed triglycerides, and consequently it will take up much less glucose from the blood.

It may seem as if it would be beneficial to have muscle cells taking up fats in the bloodstream and using them, rather than having the fats deposited in adipose tissue for long-term storage. In reality, when a muscle is utilizing fats for energy, it is using less glucose and removing less glucose from the blood. If muscle tissue is not active, as is often the case in a sedentary person, it becomes resistant to insulin and consequently takes in less glucose (Booth et al. 2008). This condition, which is characterized by hyperglycemia (high blood sugar), hyperinsulinemia (high blood insulin), and insulin resistance (tissues not responding to insulin in the blood), is known as metabolic syndrome or

TABLE 14.3.

Features of metabolic syndrome (or syndrome X)

Condition	Metabolic causes and consequences
Hyperlipidemia	Increased serum levels of triglycerides, especially VLDL.
	Triglycerides taken up by muscle tissue and utilized for energy.
	Muscle demand for glucose decreases.
Hyperglycemia	High blood sugar due to low energy expenditure and high levels of triglycerides.
	Triggers insulin release, but tissues have no capacity to absorb glucose.
Hyperinsulinemia	High blood sugar causes insulin release from pancreas.
	Low energy expenditure and high triglyceride availability decrease demand for glucose.
	Blood glucose remains high, triggering further release of insulin.
Insulin resistance	Low energy expenditure, high availability of serum triglycerides, and low demand for glucose make cells unresponsive to insulin.

Source: Information in this table summarizes discussion in the text.

syndrome X (Zammit et al. 2001). Even if the hyperglycemia and insulin resistance have not yet reached an advanced stage, they are usually precursors to type 2 diabetes mellitus. This syndrome is also associated with increased risk of hypertension and heart disease (see table 14.3).

A scenario that might lead to metabolic syndrome seems to be prevalent in a typical American setting. People with sedentary lifestyles are not getting much exercise to cause muscles to expend energy and metabolize glucose and fats. These individuals, whether adults at home, working in an office, driving a vehicle to visit clients, or adolescents in school, may be consuming snacks and sweetened beverages constantly throughout the day. If between-meal snacking happens only occasionally, the tissues may absorb and use glucose derived from the snack, because they may not have adapted to a constant or frequent presence of elevated levels of glucose in the blood and the resulting insulin secretion. Also, an occasional high-fructose soft drink will probably not be immediately funneled into fat production if there are other energy requirements in the body.

However, if snacking and soft drinks are standard fare on a daily basis in the absence of exercise, energy supplies in various tissues, particularly the liver and skeletal muscle, will not have an opportunity to become diminished. Such conditions can lead to a detrimental biochemical adaptation.

A look at the other end of the metabolic spectrum, which may be more familiar to exercise enthusiasts, may help to illustrate the body's capacity for metabolic adaptation. Marathon runners have to contend with a limited reserve of glycogen—the starch or storage form of glucose stored in muscle tissue and liver. A well-trained athlete has adapted to frequent depletion of glycogen stores by producing enzymes in those tissues (muscle in particular) that can result in more glycogen getting stored. Clever athletes often push this adaptation to the limit by a practice known as glycogen loading. This entails doing vigorous exercise a couple of days prior to a competition, which depletes glycogen stores that are already larger than normal because of constant training. Then the athlete often eats a larger than normal carbohydrate-rich meal, which results in large amounts of glucose entering the bloodstream. The glycogen-depleted tissues absorb this glucose and produce even more glycogen from it than they had before, which will be beneficial for prolonged expenditure of energy in a marathon or triathalon race (Sharman 1981). There have not been thorough studies to determine how well this technique works and how much individual variation there is, but many athletes swear by it.

Meanwhile, the sedentary individual who is constantly snacking does not allow the body to deplete its stores of energy. That person's skeletal muscles store less glycogen than they would if the muscles were being exercised regularly. If fructose is part of the constant snacking fare, either from sucrose or from high-fructose corn syrup, the sedentary body converts most of the fructose to fat. The liver sends fat into the blood for muscle and adipose tissues to use, hence serum triglycerides increase. Because inactive muscle has a relatively low energy demand, it does not respond to insulin in the bloodstream and takes up very little glucose. Consequently, glucose levels remain elevated in the blood for a longer time, which triggers continuous release of insulin. Muscle tissue will eventually adapt to this nutritional overabundance and will revert to an insulin-responsive state only when exercise is resumed.

Just walking for an hour a day will consume enough energy to stimulate more glycogen production in muscles. This forces muscle tissue to take up enough glucose to replenish glycogen stores. Consequently, muscle cells continue to respond to insulin. Skeletal muscle is the most responsive tissue to insulin, and exercising muscles can consume relatively large amounts of glucose. In addition, skeletal muscle is the most substantial tissue in a lean body in terms of mass, so it is easy to see the importance of even moderate exercise in order to avoid negative metabolic sequelae that can ultimately lead to health complications and the dreaded metabolic syndrome or type 2 diabetes.

Natural Sweeteners

Honey, a natural sweetener that contains approximately equal amounts of glucose and fructose, might be expected to cause the same metabolic complications as sucrose and high-fructose corn syrup, which also contain both glucose and fructose. However, a study in France showed that substituting honey for refined sugar in the diet of rats resulted in significantly lower levels of plasma triglycerides and higher levels of plasma vitamin E (Busserolles et al. 2002). Diets were matched for fructose content, so these results indicate there are components in honey that protect the body from the ravages of fructose. In addition, the study showed that heart muscle was less susceptible to lipid peroxidation (destruction of cell membranes) when honey was substituted for refined sugars. Higher levels of vitamin E in plasma would account for this decreased lipid peroxidation. The control diet in this study contained starch as the sole source of carbohydrate, which is digested to form glucose. The control diet produced the lowest levels of plasma triglycerides. The levels of plasma vitamin E and susceptibility of heart muscle to lipid peroxidation were about the same whether the rats were fed starch or honey diets. In general, starches do not have detrimental health effects, indicating that fructose is the major culprit when it comes to complications from dietary sugars.

Since fructose is found in fruits and fruit juices, it is natural to ask whether this source of fructose is also detrimental to health. The amount of fructose in fruits can be high but, like honey, fruits contain antioxidants or phytochemicals that suppress free-radical activity generated by fructose. The levels of sugar that occur naturally in juices (without added sugar) are generally much lower than those in soft drinks, and there seems to be a tendency toward moderation when drinking fruit juices compared with artificially sweetened beverages.

Many studies have found that starches can cause larger fluctuations in blood glucose, ergo insulin, compared with sucrose. However, this is not a consistent finding, and there appear to be many factors (including type of starch and fiber content of foods) involved in determining how large the glucose and insulin fluctuations will be (Daly 2003). There were early concerns that large fluctuations in glucose and insulin were not healthy, although the tide seems to be shifting. Smaller changes may now be considered unhealthy, particularly when there are high fasting levels of glucose and insulin. Figure 14.2 summarizes the effects that lack of exercise and high levels of dietary fructose can have on several metabolic parameters.

Obesity Hormones

Although genetic factors and endocrine dysfunction have been suspected as a cause of obesity for a long time, it was not until the late 1970s that a genetically obese strain of mice was developed by inbreeding (Coleman 1978). That strain, known as *ob/ob* mice, was found to have a defect in the gene that codes for a protein named leptin (Zhang et al. 1994). The name was derived from the Greek word *leptos*, meaning thin. It was found that leptin was expressed or synthesized predominantly in adipose tissue or fat cells. As a result of their defective gene, these mice did not produce a functional leptin protein. They ate more food when it was available, expended less energy, and became obese.

Paradoxically, it was discovered that obese humans produce more leptin than their thin counterparts. So, if a functional leptin is not produced, an individual becomes obese, as in this strain of mice. But individuals with normal leptin genes produce more leptin if they are obese than if they are thin. The higher levels of leptin in obese individuals may be due to leptin insensitivity or leptin resistance. This is analogous to type 2 diabetes, in which individuals produce insulin and have higher levels of insulin circulating in the blood, but it is not resulting in a reduction of blood glucose. The insulin receptors and their associated glucose transporters are not responding.

The body seems to have a regulatory circuit to maintain body weight and appetite. Leptin is part of this circuit, much as a thermostat is part of a circuit to control heat in a building. If there is no thermostat (or leptin), there is no control. If there is a thermostat, it can be set to either a low level or a high level and will usually maintain the temperature at the level where it is set. In this context, leptin can be viewed as a hormone that provides the body with an index of its own nutritional status—a large amount of body fat produces a higher level of leptin. However, this view seems to be falling out of favor as we learn more about how leptin is involved in the satiety response.

Another strain of mice was developed at the same time as the *ob/ob* strain, and this other strain had a defect in the gene that codes for the leptin receptor

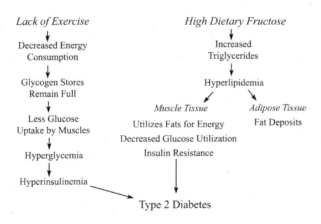

FIGURE 14.2. Metabolic consequences of lack of exercise and high sugar (fructose) consumption.

(Coleman 1978). These are known as *db/db* mice (the *db* abbreviation indicates that these mice are diabetic). It was mentioned earlier that obesity can predispose individuals to a wide range of health complications, including insulin resistance, diabetes, and metabolic syndrome. Inducing mutations in the gene for the leptin receptor in mice (receptor knockout mice) resulted in obese mice with elevated levels of leptin, insulin, glucose, and corticosterone in the blood, indicating the interrelationship of the leptin system with the glucose-insulin system (Cohen et al. 2001).

Blood levels of leptin increase with body weight and fall when a person loses weight (Maffei et al. 1995). In addition, genetically obese (*ob/ob*) mice (lacking functional leptin) lose body fat and appetite when they are injected with leptin, whereas *db/db* mice that lack the leptin receptor are not affected by leptin injections. It is possible to get both strains of mice to lose weight by limiting their food consumption, although liver pathology does not change much with food restriction. They still have fatty liver with large deposits of triglycerides, which are associated with leptin deficiency or nonfunctional leptin receptors.

The leptin system mediates its effects through control centers in the hypothalamus of the midbrain that regulate feeding behavior and appetite, as well as affecting body temperature and energy expenditure. There are several neuropeptide transmitters in the hypothalamus and throughout the body, particularly in the gastrointestinal tract. They include neuropeptide Y (NPY), cholecystekinin (CCK), agouti-related peptide (AgRP), alpha- and beta-melanocyte stimulating hormones (α- and β-MSH), obestatin, ghrelin (for growth hormone–releasing protein), beta-endorphin (endogenous morphine), and endocannabinoids (endogenous substances mimicked by *Cannabis* or marijuana) to name a few. These hormones and neurotransmitters are involved in either increasing appetite or suppressing appetite. Some affect gastric motility, or movement of food through the gastrointestinal tract. The coalition of adipose tissue, gut, bloodstream, and brain seem to be in close communication with one another to monitor the body's nutritional status and adjust the individual's behavior to eat or abstain, depending on signals transmitted.

The hormone connections in obesity are clearly complex, and there are numerous sites where a genetic variation or genetic flaw can result in an imbalance and lead to obesity. Attempts to reduce obesity by giving injections of leptin have had very little success (only about ten successful cases described) (Cohen and Friedman 2004), primarily because a defect in the gene for producing leptin is probably rare. In general, the highest leptin levels are found in the most obese individuals. It is unlikely that the recent, sudden rise in obesity in North America is due to mutations in genes that are involved in appetite and weight control. (On the other hand, the relatively constant level of extreme obesity observed in less than 1 percent of the population prior to 1980 may well have its origins in various genetic defects. There are many sites where a defective gene can lead to obesity.)

It seems more likely that the recent obesity epidemic has been triggered by changes in diet. Portion sizes are important, but sugar, specifically the fructose component of sugar, is a prime candidate. Fructose is the factor that has increased in greatest proportion in American diets in the past two decades. Furthermore, the propensity for fructose to cause glycation of proteins may add an additional layer of complexity to the diet-hormone-obesity interactions. There will be more on this in the next chapter.

The Role of Exercise Cannot Be Ignored

Exercise offers important health benefits, whether a person is obese or just overweight; its benefits also cannot be ruled out for those individuals in the ideal weight category. If the genes are not working in an individual's favor, diet and exercise become much more critical for good health. The underlying physiological and metabolic justifications for the effects of exercise on health are beginning to be understood. There is strong evidence that community-based exercise intervention programs improve health and reduce incidences of several diseases, including coronary heart disease, stroke, type 2 diabetes, breast cancer, and colorectal cancer (Roux et al. 2008).

Lifestyles have changed dramatically over the past fifty years in developed countries, particularly in the United States, and are changing rapidly in developing countries. Although opportunities for exercise are abundant, the populace tends to opt for less strenuous alternatives whenever possible. There are sometimes barriers to exercise in our daily routines. Stairwells in buildings are often locked, forcing people to use elevators rather than walking one or two flights of stairs. Many suburban neighborhoods do not even have sidewalks, forcing residents to walk in the street or drive to a local park or fitness facility. One of the biggest pastimes for retired men (the population at greatest risk for heart disease) is golf, but many golf courses today require patrons to use motorized golf carts, which results in a lost opportunity for health-conscious golfers.

Conclusions

Obesity, in and of itself, is not always life threatening. Genetics plays an important role in longevity, but diet is also an important factor. There is good evidence that increased portion sizes and between-meal snacking promote obesity. Dietary fructose, whether from sucrose or high-fructose corn syrup, is likely playing a role in many of the health complications that are frequently associated with obesity and metabolic syndrome.

There is much controversy over whether sugar is in fact addictive, in the way that certain drugs are known to be addictive. However, as researchers continue to tease the underlying physiological factors out of the complex hunger-satiety responses and desire for sweets, it becomes clear that highly sweetened foods and beverages are having an impact on the craving for more. That fructose bypasses the insulin response and tends to be converted directly to fat is well established. The insulin response, which is triggered by glucose but not by fructose, interacts with a wide range of other hormones to elicit a feeling of satiety and suppresses the appetite. The details of these interactions are only beginning to be understood. The connections between high intake of fructose from any source and insulin resistance or type 2 diabetes are compelling. People need to understand that excess sugar in the diet is driving the deposition of fat in adipose tissue.

Dietary fructose elevates serum levels of triglycerides and very low-density lipoprotein (VLDL), which increases risk of heart disease and stroke. Excess fructose in the diet arising from the high level of sweeteners in many foods may be more detrimental than saturated fats with respect to blood lipid profiles and risk of heart disease. The fats produced in the body from excess sugar are mostly saturated. In addition to its effect on the blood lipid profile, fructose is much more reactive than glucose in promoting free-radical reactions and oxidation or modification of low-density lipoprotein (LDL), which in turn leads to atherosclerosis.

Sucrose is more stable on the store shelf and not susceptible to the chemical degradation found with fructose in high-fructose corn syrup. Fructose oxidation produces formaldehyde, which may explain the rise in asthma that followed the introduction of high-fructose corn syrup in soft drinks. Although it is not possible to prove this latter relationship, the evidence is compelling enough to suggest a need for a controlled clinical study.

15

Dietary Choices
for Comprehensive Health

Cell membranes were once thought to be relatively static barriers separating a cell's surroundings from its inner workings. Today we know cell membranes are bustling with activity. Long-chain polyunsaturated fatty acids are vital membrane components that serve in cellular communication. These fatty acids are released from membranes when certain signals are received by a cell and are shuttled along metabolic pathways to form a bewildering array of potent bioactive signaling molecules. The specific agent formed from these polyunsaturated fatty acids depends on the cell that makes it—enzymes expressed in one type of cell differ from those expressed in other cells. The products formed will also depend on the essential fatty acid status of active membrane components and the types of signals received.

These mercurial lipid messengers can easily pass through membranes, so there is no mechanism for confining them in vesicular bundles (which is generally the case for nonlipid hormones and neurotransmitters) until the time is right for their release. The lipid messengers from fatty acids are formed instantaneously and migrate only short distances to elicit their actions on neighboring cells or within the cell that forms them. They exist for a brief moment to accomplish their purpose and are rapidly inactivated by enzymes that keep them in check. Their fleeting existence has made it difficult to study these elusive characters, which are often major players in the orchestration of life's complex processes.

The polyunsaturated fatty acids may also undergo spontaneous oxidation that is not dependent on enzymes. This process is known as lipid peroxidation and may destroy cell membranes, which can lead to the ultimate demise of a cell. A variety of potentially toxic and carcinogenic substances that are produced during lipid peroxidation may spread to other cells and cause persistent damage there. Although the polyunsaturated fatty acids are essential to life, it may be possible to get too much of them.

The saturated and monounsaturated fatty acids can be synthesized in the body from other nutrients, typically from excess dietary carbohydrate (sugars and starches). Consequently they are not essential fatty acids. The saturated and monounsaturated fatty acids are resistant to lipid peroxidation, and the messengers formed from them are far fewer and less formidable in their actions.

What Fatty Acids Are Better in the Diet?

There is widespread opinion that saturated fats in the diet are unhealthy. This belief stems from early studies of the effects of dietary fats on serum cholesterol in humans, done in the 1950s and early 1960s (discussed in chapter 1). The early studies used saturated fat sources that had negligible amounts of polyunsaturated fatty acids and consequently caused serum cholesterol levels to rise. Such diets would not normally be consumed in a free-living population. Today we know that moderate amounts of polyunsaturated fatty acids in the diet regulate expression of specific genes to suppress total serum cholesterol and regulate other metabolic pathways. The level of dietary polyunsaturated oils needed to accomplish a more favorable lipid profile is not certain, because genetic variation is a major influence on serum cholesterol levels in any individual. Recent research has helped to clarify some of these diverse effects of polyunsaturated fatty acids (Bordoni et al. 2006). Some studies of the impact of dietary fats on serum cholesterol are at odds with early studies that established the saturated fat–serum cholesterol association and the relationship of both to cardiovascular disease.

Two separate investigations monitored the effects of diets containing different amounts of monounsaturated versus saturated fatty acids on serum cholesterol levels in young, healthy volunteers with normal serum cholesterol levels (Ng et al. 1992; Choudhury, Tan, and Truswell 1995). The aim was to compare diets that had predominantly monounsaturated fatty acids from olive oil with diets that had nearly equal proportions of monounsaturated and saturated fatty acids from palm oil. Each of the diets had about 10 percent polyunsaturated linoleic acid. Both studies found nearly identical levels of total serum cholesterol and LDL-cholesterol in subjects at the end of thirty days or six weeks on each experimental diet. These results refute the notion that a higher ratio of saturated relative to monounsaturated fatty acids in dietary fats produces a less favorable lipid profile that might increase the risk of heart disease.

Another study compared diets containing palm oil with high-oleic sunflower oil as the major sources of fats and showed a significant elevation in serum total cholesterol and LDL-cholesterol of subjects when they consumed the palm oil diet compared with the high-oleic sunflower oil diet (Denke and Grundy 1992). This latter study used older male subjects in a Veterans Administration Hospital, and more than one-third of the participants had a history of coronary heart disease. Whether the different response was due to age, exclusively male subjects, health status, the composition of the liquid-formula diet, or other factors is difficult to say. The high-oleic sunflower oil also contained 58 percent more linoleic acid than the palm oil, which would be a significant factor since palm oil has less than 10 percent linoleic acid. Perhaps the older subjects confined to the metabolic ward of a hospital had a genetic predisposition for the adverse response to changes in dietary fats. An important point is that some people do not have an adverse response to dietary saturated fats relative to unsaturated fats, while others may experience an unfavorable change in their serum lipid profile when similar changes in diet are administered.

As these and other studies demonstrate, recommendations regarding what dietary fats are beneficial and what ones should be avoided are complicated by the diverse profile of fatty acids in the foods we eat and the heterogeneous genetic makeup of the population. The assortment of enzymatic reactions some of these fatty acids undergo as a result of the genetic dissimilarity adds to the complexity. The polyunsaturated fatty acids help maintain the normal function of various physiological systems, in addition to their involvement in the regulation of cholesterol synthesis. Because of the complexity of essential fatty acid metabolism and function, it is not easy to come up with a simple remedy for each situation. A dietary recommendation that may benefit one ailment could exacerbate another. It is also important to consider which dietary manipulations will be easy to implement and follow over a long period of time—perhaps a lifetime.

There is a widely held belief that lower total fat and less saturated fat in the diet is healthier. This is based on certain assumptions about dietary fat that are beginning to falter under the scrutiny of controlled scientific experiments, especially with respect to obesity and its associated adverse health effects, which are commonly known as metabolic syndrome.

Guidelines for Weight Loss

The incidence of obesity has been on the rise since the 1980s in the United States and is considered by many to have reached epidemic proportions. Obesity increases the risk of many diseases, as discussed in the previous chapter. Most notable among health complications are insulin resistance, type 2 diabetes, and high blood pressure, which all increase the risk of heart disease.

The key to effective weight loss is to consume less energy than you expend. There are essentially two factors that need to be adjusted to accomplish this. One is to eat fewer calories, and the other is to exercise more to burn energy. A combination of these two practices is generally the best approach. How one chooses to exercise more will vary with the individual. It is best to set up an exercise program that can be sustained throughout life. Because the exercise component will vary so much from person to person, it is better that each individual get specific advice tailored to his or her goals and capabilities. It is important to achieve a regular exercise regimen that is enjoyable and will therefore become a permanent routine. Exercise is important not only to lose weight but also to maintain or attain general good health.

In order to decrease energy intake, a reduction of fat in the diet would seem like the most obvious approach, since fats supply more than twice as many calories as carbohydrates or protein on a gram for gram basis. However, there are certain considerations that must be taken into account when modifying the diet for long-term weight loss. First and foremost is that total calorie intake must be reduced, and the best way to accomplish this is to eat more satisfying foods but smaller quantities of them. A low-fat diet generally has a higher level of energy (calories) from carbohydrates. There are problems with this approach, and studies have shown that it is usually not successful over an extended period of time. An analysis of low-fat and calorie-restricted diets for weight reduction showed there was not a significant difference in the two approaches (Pirozzo et al. 2002).

In the early 1970s Robert Atkins developed what is widely known as the Atkins diet (Atkins 1972). It is a low-carbohydrate, high-fat, and high-protein diet. A few notable clinical trials compared low-fat and calorie-restricted diets with low-carbohydrate, high-fat diets in obese patients. Patients on the low-carbo-hydrate, high-fat diet were not restricted in their total calorie intake. In other words, they were free to eat as much as they wished to feel satisfied, whereas patients on the low-fat diets were strongly encouraged to carefully regulate their total calorie intake. These randomized, controlled trials showed that subjects consuming low-carbohydrate, high-fat diets lost more weight over the course of six months than those on calorie-restricted low-fat diets (Brehm et al. 2003; Samaha et al. 2003; Yancy et al. 2004). The low-fat, calorie-restricted regimen in one of these studies was based on obesity management guidelines from the National Heart, Lung, and Blood Institute (Samaha et al. 2003). Subjects on the low-carbohydrate, high-fat diet in that study also showed greater improvement in serum triglycerides and greater insulin sensitivity in participants without diabetes relative to subjects on the low-fat diet (Samaha et al. 2003).

A more recent trial compared four popular dietary approaches for weight loss and found the Atkins high-fat diet produced the greatest weight loss over a period of one year (Gardner et al. 2007). The high-fat diet also showed signifi-cant improvements in several health risk parameters compared with some or all of the other diets, including lower serum triglycerides, higher HDL-cholesterol, and lower blood pressure. In all these trials the high-fat, low-carbohydrate diets did not result in any significant change in LDL-cholesterol, but HDL-choles-terol levels were improved in some. The results of these studies suggest that guidelines for a prudent diet set by the National Institutes of Health and other organizations, such as the U.S. Department of Agriculture and the American

Heart Association, particularly for management of obesity, may need to be reconsidered.

How the High-Fat, Low-Carbohydrate Diet Works

There is a physiological explanation for the seemingly unexpected success of the high-fat, low-carbohydrate diet. The fructose component of sugar or high-fructose corn syrup raises serum triglycerides, primarily in very low-density lipoprotein (VLDL). Absorbed glucose from these sweeteners and from starch causes insulin secretion, which in turn signals cells to remove glucose from the circulation. The majority of glucose is taken up by muscle, liver, and adipose tissue in a normal, insulin-responsive individual. Insulin signals set in motion a metabolic shift toward storage of energy, either as glycogen (starch) in muscle and liver, or as triglycerides in adipose tissue. When glycogen stores are filled in the liver, triglycerides are formed and packaged into VLDL particles for distribution to muscle and adipose tissue. Muscle tissue uses triglycerides for energy, whereas adipose tissue stores triglycerides as fat. Higher levels of triglycerides and VLDL in blood also seem to correlate with small, dense low-density lipoprotein (LDL) particles, which are known to be a greater risk factor for heart disease than larger LDL particles (Samaha, Foster, and Makris 2007).

High-carbohydrate consumption can cause blood glucose levels to rise faster and to higher levels than would occur with lower amounts of carbohydrate in a meal. This causes more insulin to be secreted, which triggers a more rapid decrease in blood glucose that ultimately overshoots the normal resting glucose level. Consequently, another metabolic shift is initiated to bring glucose levels back up to normal. This latter response is accompanied by a variety of hormonal signals, as discussed in the previous chapter, that ultimately result in a hunger sensation and a desire to eat again within a couple of hours.

When there is less carbohydrate in the form of sugars and starches in the diet, blood glucose levels will not rise as much and therefore less insulin will be secreted. In addition, if there is a significant amount of fiber in the diet, it will bind sugars and starches and slow their digestion and absorption from the gastrointestinal tract. This results in less dramatic fluctuations in glucose and insulin and a slower return to normal without going much below the resting glucose level. Consequently, fiber in a meal helps to delay the hunger pangs often associated with high-carbohydrate diets that are low in fiber, as one finds with many sweetened and starchy foods. An individual consuming a low-carbohydrate diet, which generally means it is higher in fat, will experience less frequent desire to eat and will naturally consume fewer calories. The studies mentioned above did not force participants on the high-fat diets to restrict their food intake; the participants had a longer duration of satisfaction and ate less because they were not hungry as often. On the other hand, subjects on the low-fat, high-carbohydrate diets were forced to restrict their caloric intake because those diets spark a frequent desire to eat.

The low-fat, high-carbohydrate diet, as a result of stimulating insulin secretion, also suppresses fatty acid oxidation for energy metabolism in muscle tissue in response to insulin signaling mechanisms (Wood 2006). Patients with insulin resistance would benefit from a high-fat, low-carbohydrate diet because there would be less stimulation of insulin release owing to less glucose being absorbed from the intestinal tract. There would also be less sugar available for synthesis of fatty acids and triglycerides by the liver and subsequent distribution via the blood in VLDL particles for storage in adipose tissue. Muscle tissue will also utilize more fat when there is less insulin in the circulation because the glucose transporter in muscle cells will be less active.

The high-fat diet approach to losing weight runs counter to popular belief, but it is easily explained by well-understood physiological principles and supported by hard scientific evidence. It is not going to be easy to persuade some policy makers to reconsider the low-fat diet recommendations, although many organizations, including the American Diabetes Association, are beginning to recognize the merits of the low-carbohydrate dietary approach. Perhaps the most difficult thing to accept is that lowering the proportion of carbohydrates in a 'diet means that other nutrients are necessarily increasing in proportion, and fat is the most likely candidate.

Protein has not entered much of this discussion mainly because it is generally around 15 percent of total energy in a diet, which for most purposes is suitable. There would generally be no problem if protein intake increased as a result of lowering dietary carbohydrate. However, very high levels of protein in a diet can be risky because of the need to eliminate nitrogen when amino acids are metabolized for energy. There could be a dangerous buildup of ammonia in some individuals. Ingesting dangerously high levels of protein can be accomplished only by consuming purified protein and/or hydrolyzed protein supplements. Natural foods do not have such high levels of protein.

Nutrition Facts on the Label

A nutrition facts table is now found on labels for most packaged foods sold in the United States and in many other countries. Certain information is required on these labels, while other information is optional. The top of the nutrition facts table indicates the serving size and number of servings per package. The serving size is important, since packages that are normally eaten as a single portion are often designated as more than one serving. The information in the table is generally for a single serving: the number of calories (kilocalories), for example, is given per serving. If there are multiple servings per package, then the total calories in the package must be obtained by multiplying the calories per serving by the number of servings per package. The number of calories listed on the label can be deceptive when there are two or more servings in the container. Labels typically also include calories from fat. The remainder of calories (those not from fat) would come from carbohydrates and protein.

Topping the list of specific nutrients is fat, with the amount of total fat, saturated fat, and trans fat given in grams for each. The amount of unsaturated fat can be calculated by adding the amounts of saturated and trans fats and subtracting that sum from the total fat. However, this does not give a breakdown for the amount of polyunsaturated fatty acids versus monounsaturated fatty acids. The amount of unsaturated fat is sometimes listed, particularly when the manufacturer wants to promote the product because of a certain fatty acid profile. As I have stressed, the amount of omega-6 polyunsaturated oils in foods should be of greater concern to many individuals than the amount of saturated fats because the omega-6 fatty acids can promote inflammation and a variety of other syndromes.

There is rarely a breakdown of the amount of omega-3 and omega-6 fatty acids in the nutrition facts table. This information would be good to have, since we should try to improve the proportion of omega-3 fatty acids in the diet. However, there are relatively few significant sources of omega-3 fatty acids in vegetable products, and fish oils are not added to food because they are easily oxidized and become rancid. In some cases algal extracts are used to boost the omega-3 content of foods for promotional purposes. Since the relative amounts of omega-3 versus omega-6 polyunsaturated fatty acids are important to know for consumers concerned about inflammation problems, it may be time to reevaluate what types of fats are listed in the nutrition facts table.

Meats and products made from animal fats (often dairy products) are a major source of saturated fat in the diet. However, there are other factors in meats that can contribute to adverse health effects, as mentioned in earlier chapters. Processed meats generally have preservatives such as nitrites to improve shelf life and give a better appearance. Preservatives in processed meats could be the underlying cause for the widespread association of saturated fats with bad health, as discussed in chapter 10 (Norat et al. 2005). It would be valuable for future studies of relationships between diets and diseases to make greater distinctions with regard to other factors found in foods that are generally classified as high in saturated fats.

The requirement to list trans fat on the label has resulted in food processors drastically reducing the amount of trans fat in foods. Although the amount is not always zero as many labels imply, the amount is much less than it was in the 1980s. If the amount of trans fat per serving is less than one-half gram, it can be listed as zero. This can be deceptive for many products. Under current rules, if a typical serving has five grams of total fat, nearly 10 percent of that can be trans fat and yet the trans fat content of the food product can still be listed as zero grams. In order to reduce amounts of trans fat, the food industry has reformulated many foods to contain some fully hydrogenated vegetable oils, which are saturated fats. They combine these with regular vegetable oils, which are mostly polyunsaturated and monounsaturated fatty acids. With such mixtures the final products have a suitable consistency while containing lesser amounts of the partially hydrogenated vegetable oils, which contain trans fats.

The amount of cholesterol is also listed; it is zero for vegetable-derived foods and present at some level when there is animal fat in the food. The amount of cholesterol in a product should generally be of little concern, since most foods contain relatively small amounts of cholesterol and those that have larger amounts are not normally eaten in large quantities. Consuming low levels of polyunsaturated fatty acids will have a greater impact on lowering serum cholesterol levels than consuming moderate amounts of dietary cholesterol would have in raising serum cholesterol. A typical diet generally has sufficient levels of polyunsaturated fatty acids to give nearly optimal suppression of serum total cholesterol and LDL-cholesterol. The total intake of cholesterol in a day should be less than about 600 to 800 mg, an amount that will not significantly affect serum cholesterol levels. Eggs are a major source of cholesterol in many diets, with essentially all the cholesterol in the yolk, and are sometimes avoided for that reason. Yet even consuming as many as two eggs per day would not raise serum cholesterol noticeably, and egg whites are a good source of protein.

Sodium is generally listed after cholesterol, and patients with high blood pressure or certain electrolyte imbalances are generally advised to limit their salt and sodium intake. Most people do not have a problem handling moderate dietary amounts of sodium and normally excrete any excess sodium in urine and sweat.

Protein is generally the last nutrient listed in the nutrition facts table, but I will mention it here before carbohydrates. There has not been much said about protein in this book, and there are many different ideas about whether we are getting enough or too much protein in a typical diet. It should suffice to say that the typical American diet has ample amounts of protein and a variety of different proteins to avoid specific amino acid deficiencies.

Following the discussion regarding problems associated with high-carbo-hydrate, low-fat diets, it should be clear that the carbohydrate section of the nutrition facts table is most important. The term carbohydrate includes simple sugars, starches, and fiber. Fiber does not contribute to the caloric content of carbohydrates in foods, because it is not digested and not absorbed. The term digestible fiber is often used, which is a misnomer in human nutrition because it is not digested but is soluble in water. Fiber is beneficial in slowing absorption of sugars from the intestine, as mentioned above, but is also important in helping to flush waste products from the intestine. Fiber also adds bulk to the feces for better intestinal function. The larger the portion of dietary fiber in one's total carbohydrate intake the better, since fiber does not contribute to total energy intake and moderates the absorption of glucose from the breakdown of other carbohydrates.

This brings us to the sugar entry, which needs the most attention from anyone striving for a healthy diet. Sugar intake should be as low as possible, whether one is attempting to lose weight or to minimize the risks of developing many diseases. Sugars are chemically reactive and can promote production of free radicals and toxic by-products. Fructose is many times more reactive than glucose in generating free radicals, cross-linking proteins, and producing toxins. High-fructose corn syrup has a higher proportion of the more reactive fructose than table sugar (sucrose), and the fructose in high-fructose corn syrup is free to react and form toxic products such as formaldehyde when standing in solution on a shelf, as is the case for many soft drinks. Although honey is a mixture of nearly equal portions of glucose and fructose, it contains antioxidants that slow the decomposition of fructose while standing and was shown to have less detrimental effects on several parameters with regard to overall health (Busserolles et al. 2002).

The amount of sugar in many common food products is quite astounding. Sugars (sucrose and high-fructose corn syrup) are added to make foods more desirable. As mentioned in the previous chapter, the per capita consumption of sweeteners rose more than 20 percent between the mid-1980s and 1998. Perhaps a major national initiative to educate the public about the adverse health effects of sugars needs to be launched. An approach similar to the one used forty or fifty years ago regarding the hazards of a diet high in saturated fats and cholesterol would be a good way to start. That program was quite successful in making people conscious of the issue. The public may not have heeded the advice in a manner that many health authorities would have liked to see, but the initiative certainly raised awareness and caused major changes. Whether that dietary advice was correct or effective for reaching the intended goal is also debatable, but the point is that the public education program brought attention to the issue.

Dietary sugar is a graver health issue at this time than dietary fat, and the scientific evidence for the consequences of excessive sugar intake is even more compelling than the evidence was in the 1960s for an association between saturated fat and high serum cholesterol. There is a strong correlation between the rising incidence of obesity in America, with its associated metabolic complications, and the rise in per capita sugar consumption (Elliott et al. 2002). The association of high serum cholesterol with heart disease led to the assumption that saturated fats were unhealthy. Today a more thorough scientific understanding of the processes involved in atherosclerosis and heart disease should shift the implications away from saturated fats with regard to risk of life- threatening disease. Sugars may have a concealed role in the health effects attributed to saturated fats, since some studies showed an intercorrelation between dietary saturated fats and sugar, such as the Seven Countries Study mentioned in chapter 1 (Keys 1980, 252).

Rating the Fats and Oils

Since oils are a major source of fats in the diet, many consumers may want to know which oils are best to use with regard to overall health. Olive oil is good because it has mostly monounsaturated fatty acids and about 10 percent polyunsaturated fatty acids. Palm oil, which was denounced years ago because it has nearly 50 percent saturated fatty acids, would also be a good choice, with nearly 50 percent monounsaturated and about 10 percent polyunsaturated fatty acids, which would be sufficient to suppress cholesterol production. Other popular oils are often considered to be good because of their high monounsaturated fatty acid content, but they have higher amounts of the omega-6 linoleic acid than olive or palm oil. Canola oil and peanut oil have about 30 percent polyunsaturated fatty acids, although canola oil normally has about 10 percent omega-3 linolenic acid in the polyunsaturated fatty acid fraction. A table of the fatty acid composition of common dietary fats and oils can be found in appendix B.

The agricultural industry has developed seed oils with high oleic acid content. These include high-oleic versions of canola oil, safflower oil, and sunflower oil. High-oleic sunflower oil has only about 11 percent linoleic acid, and high-oleic safflower oil has about 13 percent linoleic acid. High-oleic canola oil has about 17 percent linoleic acid but has 2 percent omega-3 linolenic acid to offset the higher linoleic acid content in terms of potential physiological effects. In general, a lower polyunsaturated fatty acid content for a vegetable oil makes it more suitable for frying. Fried foods should be kept to a minimum in the diet because of adverse chemical reactions that can occur at high temperatures in most oils. Saturated fats would be the most suitable and least susceptible to oxidative breakdown at high temperatures, even though saturated fats have received bad press through the years.

Oils that are most suitable for salad dressings are different from those most suitable for frying or baking. The arguments above regarding low amounts of omega-6 polyunsaturated fatty acids in oils apply to salad dressings. Olive oil is generally considered optimal for this purpose because of its fatty acid profile and taste. Other suitable oils for salad dressings include high-oleic sunflower oil and high-oleic safflower oil as well as palm oil, where these are available. There are many oils available now in most supermarkets that have better polyunsaturated fatty acid profiles. Check appendix B for the fatty acid composition of many commercial oils. Consumers should be aware that mayonnaise is generally made from vegetable oils that may have a high omega-6 content and should be consumed in moderation for that reason.

The best vegetable source for omega-3 fatty acids is flaxseed oil, which has more than 50 percent linolenic acid and about 12 to 13 percent omega-6 linoleic acid. Other oils with significant amounts of omega-3 linolenic acid are walnut oil (10 percent), canola oil (9 percent), and soy oil (7 percent). Wheat germ is also a source of omega-3 linolenic acid. All these latter oils with 10 percent or less of omega-3 fatty acid have more than 50 percent linoleic acid. However, the linolenic acid content helps to suppress metabolism of the linoleic acid to bioactive omega-6 fatty acid derivatives.

Conclusions

By this time the reader may feel completely confused. After all, this book is arguing that saturated fats may not be as bad for our health as is commonly believed but polyunsaturated oils, particularly omega-6 vegetable oils, are not healthy when consumed in excess. It is also saying that low-fat diets can be worse than a high-fat diet. These ideas contradict the doctrine of the low-fat, low-saturated fat, and low-cholesterol diet that has been prescribed for decades. I feel it is time to reassess the dogma that saturated fats are bad and unsaturated fats are good. I am not calling for the pendulum to swing completely in the other direction. We need to recognize the diverse genetic constitution of the human population and understand that some people are more prone to certain health problems than to others. The aim of this book is to provide information about what diet-derived fats are doing in the body, particularly the essential fatty acids, in order to make people more aware of the impact that polyunsaturated fatty acids in vegetable oils have on our health.

We have seen how polyunsaturated fatty acids are susceptible to lipid peroxidation, which can lead to destruction of cell membranes, formation of toxic by-products that cause cancer, and deposition of low-density lipoprotein (LDL) along arteries to promote atherosclerosis. The polyunsaturated fatty acids are metabolized to a host of bioactive compounds throughout the body that are involved in signaling cells to change their course. The lipid signals may work in concert with cytokines (proteins) to stimulate the immune system in warding off infection, but overstimulation or errant signals could lead to autoimmune disease, allergic reactions, and excessive inflammation. Some lipid messengers signal platelets to aggregate and blood vessels to constrict to initiate a blood clot when we cut ourselves, but premature clotting from damaged plaque in coronary arteries causes heart attacks. Other eicosanoids derived from these essential fatty acids can promote bronchial constriction during an asthma attack and exacerbate the life-threatening symptoms.

These systems evolved in nature to protect us from a wide range of assaults in our daily lives. A shift away from saturated fats has increased the relative amount of vegetable oils and processed foods in our diets. Consequently, polyunsaturated fatty acids have become a copious presence in cell membranes. These essential fatty acids may be aggravating many physiological responses. There is evidence that the shift in proportions of omega-6 fatty acids to omega-3 fatty acids in the diet may be an underlying factor in the growing incidence of several maladies of our modern times. The pressure to decrease overall fat consumption seems to have neglected the shift toward more carbohydrates in the diet. This probably has been a major factor contributing to the obesity epidemic in the United States (Elliott et al. 2002). Although portion size is another factor, the high-carbohydrate proportions in the diet result in a shorter time period until hunger returns and more frequent food consumption.

How does one go about designing a diet that will accomplish the intended goals? Perhaps the first and foremost rule is to focus on the carbohydrates. Try to decrease the amount of sugars especially, but also pay attention to starches. Both of these contribute to the blood glucose and insulin fluctuations that drive the frequency of feeling hungry and the desire to eat. Sweetened soft drinks are a major source of sugars and should be consumed only in moderation, if at all. This is especially important for children. Current thinking would have us believe that a hamburger or other meat is the bane of our good health, with its saturated fats and artery-clogging cholesterol. The above arguments regarding the detrimental effects of high-carbohydrate diets would instead implicate the hamburger bun, side of potatoes, and sugar-laden soft drink that come with the hamburger.

We also need to consider moderation when looking at fats. This book is not arguing that we need to cut out polyunsaturated oils completely but, rather, consume them in moderation. The amounts needed to satisfy our daily essential fatty acid requirements are quite small, about 1 to 2 percent of total energy intake (Sanders 1999). Consuming as much as 5 to 6 percent of energy as essential fatty acids may be optimal and enough to suppress the production of serum cholesterol and LDL, although the amount necessary for the latter effect has not been established. When polyunsaturated vegetable oils approach 10 percent of energy in the diet, there may begin to be adverse effects in some individuals who are prone to inflammation and its associated health complications. Omega-3 fish oils in the diet can counteract some of the adverse health effects of omega-6 vegetable oils, but if there is an overwhelming surplus of omega-6 fatty acids, a small change in omega-3 fatty acids from fish or dietary supplements may not make a significant impact. When inflammatory conditions are prevalent in a patient, it is best to decrease consumption of vegetable oils, including foods such as mayonnaise, margarine, and others in which omega-6 oils are prominent ingredients. Saturated fats, monounsaturated oils, and omega-3 oils do not promote inflammation, so it should not be necessary to eliminate all fats from the diet to treat inflammatory syndromes.

The take-home message after all this is that omega-6 fatty acids can promote several unhealthy conditions. Although omega-6 vegetable oils have been shown to exacerbate certain inflammatory syndromes in animals compared with dietary saturated fats and omega-3 oils, their effects on some of the other syndromes may be less well documented. Studies on neurological disorders in humans have shown favorable results from dietary manipulation in some cases, but the underlying cause of many of the neurological disorders discussed in the earlier chapters are not that well understood. The main point is that omega-3 dietary supplements and foods with omega-3 fatty acids do not pose any particular risk in the diet with regard to health complications. But in order to reap the greatest benefit from increased omega-3 fatty acids in the diet, it may be necessary to decrease the consumption of omega-6 vegetable oils. This gives the omega-3 fatty acids a greater chance to compete with the omega-6 fatty acids for metabolism to the bioactive compounds. The omega-3 bioactive derivatives generally elicit a more moderate physiological response, which is important if the physiological system in question is overreacting or causing pain.

The story of the essential fatty acids is long and complex, but it is not completely baffling. The bioactive eicosanoids are powerful agents that can sometimes get out of control. If we have a better understanding of their functions, it will be easier to appreciate the dietary manipulations that are needed to improve our general health and well-being. The saturated fats deserve a reprieve from accusations regarding their adverse health effects. They have been associated with unhealthy conditions, but we need to look more closely at whether the culprits are saturated fats per se or whether these fats are wrongly accused because of their association with other dietary bad actors, namely the preservatives in processed meats and the sugars that accompany them in many diets.

APPENDIX A.
FUNDAMENTALS OF
CHEMICAL BONDING AND POLARITY

Chemical symbols are generally used to represent atoms of each element in diagrams of molecules; for example, H, O, N, and C represent hydrogen, oxygen, nitrogen, and carbon, respectively. The chemical nature of these elements dictates that, generally, H can have only one bond (or attachment) to other atoms, whereas O will have two bonds, N three bonds, and C four bonds, as shown in figure A.1.

Carbon atoms always have four bonds in biological molecules, but often there are double bonds to another atom, which count as two bonds. When double bonds occur between carbon atoms, the molecule is said to be unsaturated, meaning that it is not saturated with hydrogen atoms. If there are two different atoms or groups of atoms attached to each carbon forming the double bond, such as H (hydrogen) and R (any carbon group) as shown in figure A.2, it is possible to have cis or trans geometric isomers for that double bond. A cis double bond has two hydrogens on one side of the plane of the double bond and two R groups on the other side of the plane of the double bond, as seen in the center of figure A.2. A trans double bond has the H atoms across the plane of the double bond. This necessitates having the R groups also on opposite sides of the plane of the double bond, as seen on the right of figure A.2. For the

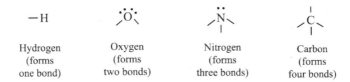

Hydrogen (forms one bond)   Oxygen (forms two bonds)   Nitrogen (forms three bonds)   Carbon (forms four bonds)

FIGURE A.1. Normal bonding in organic chemistry. Normal bonding for hydrogen is one bond, oxygen two bonds, nitrogen three bonds, and carbon four bonds. Pairs of nonbonding electrons that impart polarity to molecules containing oxygen and nitrogen are shown as two dots. Oxygen has two pairs of nonbonding electrons, and nitrogen has one pair.

purposes of the present discussion, the cis and trans designation occurs only when there is one and only one H atom on each of the carbons forming the double bond, as shown in figure A.2.

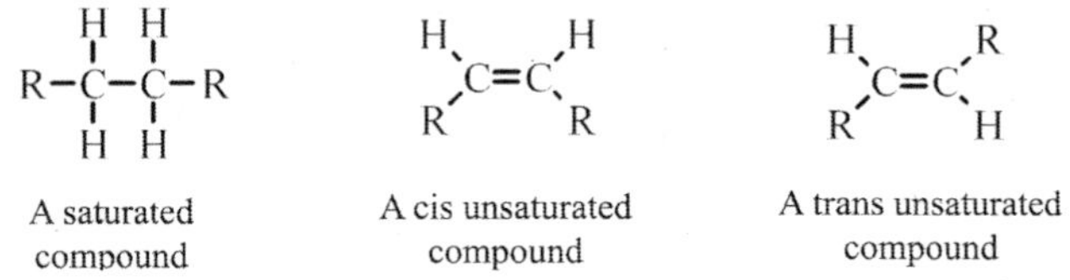

| A saturated compound | A cis unsaturated compound | A trans unsaturated compound |

FIGURE A.2. Saturated and unsaturated molecules. A saturated organic molecule (*left*), a molecule with a cis double bond (*center*), and a molecule with a trans double bond (*right*). Note the positions of the two hydrogen (H) atoms in the unsaturated compounds that determine whether the double bond is cis or trans.

The number of bonds formed by N and O atoms can differ from the norm, but there will be a charge (positive or negative) when that happens. For example, oxygen atoms are often found with only one bond (one fewer than normal) to another atom and consequently carry a negative charge, as seen on the upper right in figure A.3. Nitrogen atoms are often found with four bonds (one more than normal) and carry a positive charge, as shown on the lower right in figure A.3. The minus and plus symbols are used as a convention to indicate an imbalance or uneven distribution of electrons in the molecules, giving rise to polarity.

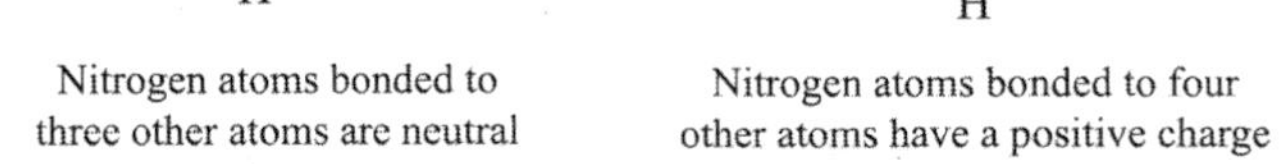

Oxygen atoms bonded to two other atoms are neutral

Oxygen atoms bonded to only one other atom have a negative charge

Nitrogen atoms bonded to three other atoms are neutral

Nitrogen atoms bonded to four other atoms have a positive charge

FIGURE A.3. Neutral molecules. Water (*top left*) and ammonia (*bottom left*) are representative of neutral molecules containing oxygen and nitrogen, respectively. It is possible for water (and certain organic molecules containing oxygen) to lose a hydrogen ion, resulting in an anion (a negative charge on the oxygen atom in that molecule). Ammonia (and organic molecules containing nitrogen) can have a hydrogen ion (H+) bind to the lone pair of electrons on nitrogen, forming a cation (a positive charge around the nitrogen atom in that molecule).

Polarity or lack of polarity in molecules arises from the nature of the chemical elements and the bonds holding the atoms together in these molecules. In biological molecules, oxygen and nitrogen atoms will generally make the molecules more polar, whereas carbon (the predominant atoms in biological molecules) will tend to make the molecules less polar. Hydrogen is the fourth common element in biomolecules, but its influence on polarity is determined by the atoms to which it is attached.

The polar molecules behave like small magnets in the sense that opposite poles attract one another and like poles repel one another. Water and ammonia molecules are shown in figure A.4, with plus and minus signs representing the positive and negative poles of the molecules. The arrows next to the molecules in that figure are pointing in the direction of the negative pole, much as the needle of a compass points toward the northern magnetic pole of the earth. This polarity makes water molecules align toward the opposite poles of their neighbors and adhere tightly together, much as a clump of magnets will do, giving rise to droplets or beads of water on a waxed (nonpolar) surface. On the other hand, water spreads out on a surface that has been worn or oxidized (a surface with many oxygen atoms attached to it, making it polar).

Nonpolar substances, such as oils, will not mix with water, because the strong attractions pulling water molecules together will exclude nonpolar substances. Therefore, the nonpolar substances will be relegated to mix with one another. A nonpolar hydrocarbon molecule is shown at the bottom of figure A.4. All electrons on the carbon atoms are involved in bonds with hydrogen atoms in this molecule, resulting in an even distribution of electrons around the atom.

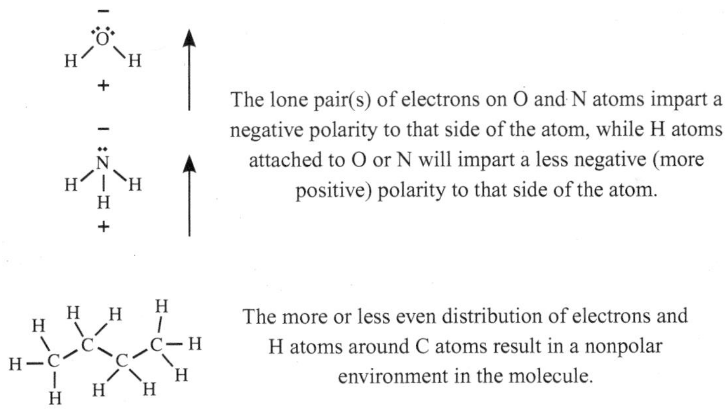

The lone pair(s) of electrons on O and N atoms impart a negative polarity to that side of the atom, while H atoms attached to O or N will impart a less negative (more positive) polarity to that side of the atom.

The more or less even distribution of electrons and H atoms around C atoms result in a nonpolar environment in the molecule.

FIGURE A.4. Molecular polarity. Representation of polarity in water (*top*) and ammonia (*middle*) with the negative pole shown by a minus sign (−) and the positive pole by a plus sign (+) for each molecule. The arrows point in the direction of the negative pole of the molecule. A nonpolar organic molecule (butane) is shown at the bottom for comparison. The organic molecule lacks polarity because of uniform distribution of electrons and charge around the carbon atoms.

Nonpolar substances are referred to as hydrophobic (water fearing), whereas substances that mix with water are called hydrophilic (water loving). Similarly, hydrophobic substances that prefer to associate with nonpolar substances such as lipids are called lipophilic (lipid loving). Many hydrophobic drugs or other substances that enter the body will tend to seek fat deposits and reside in those nonpolar environments.

APPENDIX B.
FATTY ACID COMPOSITION OF DIETARY FATS AND OILS

Several of the common fatty acids, saturated and unsaturated, are listed in table B.1 with their symbolic representation (number of carbon atoms:number of double bonds, and position of the double bonds relative to the acid group). The omega-3 or omega-6 designation of the polyunsaturated fatty acids refers to the number of carbons from the methyl end of the chain to the first double bond. Common food sources of each fatty acid and its melting point are also given. The melting point is useful for determining whether fats containing these fatty acids will tend to be liquid or solid at room temperature (about 22 to 25°C). However, triglycerides (fats) generally contain more than one type of fatty acid, so their melting points will vary depending on which fatty acids predominate.

Table B.2 is a list of some common food sources of fats and oils, with their approximate percentages of the more common fatty acids. If a fatty acid constitutes less than 1 percent, it is not listed (except for a few cases). The percentage of each fatty acid in triglycerides from any given source will vary depending on environmental and genetic factors for plant sources and dietary and genetic factors for animal sources. Some oils from genetically selected varieties of plants used for seed oils are listed. This table is not intended to be exhaustive.

TABLE B.1.

Common fatty acids, their sources, and their melting points

Common name	Symbolic representation	Common source	Melting point (°C)
Saturated fatty acids			
Butyric acid	C4:0	Butter	−7.9
Caproic acid	C6:0	Butter	−3.4
Caprylic acid	C8:0	Coconut oil	16.7
Capric acid	C10:0	Coconut oil	31.6
Lauric acid	C12:0	Coconut oil	44
Myristic acid	C14:0	Butter, coconut oil	52
Palmitic acid	C16:0	Palm oil, butter, lard	63
Stearic acid	C18:0	Beef tallow, cocoa butter	70
Monounsaturated fatty acids			
Palmitoleic acid	C16:1, Δ^9	Palm oil	−0.5
Oleic acid	C18:1, Δ^9	Canola oil, olive oil	13
Erucic acid	C22:1, Δ^{13}	Rapeseed oil	33.4
Omega-6 polyunsaturated fatty acids			
Linoleic acid	C18:2, $\Delta^{9,12}$	Soy, safflower, corn oil	−9
Dihomo-gamma-linolenic acid	C20:3, $\Delta^{8,11,14}$	Evening primrose oil	
Arachidonic acid	C20:4, $\Delta^{5,8,11,14}$	Minor constituent of foods	−49
Docosapentaenoic acid	C22:5, $\Delta^{4,7,10,13,16}$	Found in brain	
Omega-3 Polyunsaturated fatty acids			
Linolenic acid	C18:3, $\Delta^{9,12,15}$	Linseed, parilla oil	−17
Eicosapentaenoic acid	C20:5, $\Delta^{5,8,11,14,17}$	Fish oils	−53
Docosahexaenoic acid	C22:6, $\Delta^{4,7,10,13,16,19}$	Fish oils	< −50

Source: Strayer et al. 2006.

TABLE B.2.

Fatty acid composition (%) of common dietary fats and oils

	Saturated				Monounsat		Polyunsat	
	4:0–12:0	14:0	16:0	18:0	16:1	18:1	18:2	18:3
Almond oil[a]			6.5	1.7	0.6	69	29	
Avocado[a]			11	0.7	2.7	68	12.5	1
Beef tallow		3	24	19	4	43	3	1
Butterfat (milk or butter)	13[b]	11	27	12	2	29	2	1
Canola oil			4	2		62	22	10
High-oleic canola			4	2		75	17	2
Chicken fat	1	1	24	6	6	40	16	1
Cocoa butter			26	34		34	3	
Coconut oil	62[c]	18	9	3		6	2	
Corn oil			11	2		28	58	1
Cottonseed oil		1	22	3	1	19	54	1
Flaxseed oil[a]			5	4		20	13	53
Grape seed oil[a]			7	3		16	70	
Hazelnut oil[a]		0.1	5.2	2	0.2	78	10	
Lard		2	26	14	3	44	10	
Olive oil			13	3	1	71	10	
Palm oil		1	45	4		40	10	
Palm kernel oil	55[d]	16	8	3		15	2	1
Peanut oil			11	2		48	32	
Safflower oil			7	2		13	78	
High-oleic safflower			7	2		78	13	
Sesame seed oil[a]			9	5		39	41	0.3
Soybean oil			11	4		24	54	7
Sunflower oil			7	5		19	68	1
Mid-oleic sunflower			4	5		65	26	
High-oleic sunflower			4	5		79	11	
Walnut oil[a]			7	2	0.1	22	53	10
Wheat germ oil[a]			16.6	0.5	0.5	15	55	7

Sources: Strayer et al. 2006; Cordain 2009.

[a] Source = The Paleo Diet website (Cordain 2009).
[b] Butterfat and milk contain 4% butyric, 2% caproic, 1% caprylic, 3% capric, and 3% lauric acids.
[c] Coconut oil contains about 1% caproic, 8% caprylic, 6% capric, and 47% lauric acids.
[d] Palm kernel oil contains 3% caprylic, 4% capric, and 48% lauric acids.

APPENDIX C.
CHEMICAL STRUCTURES OF EICOSANOIDS

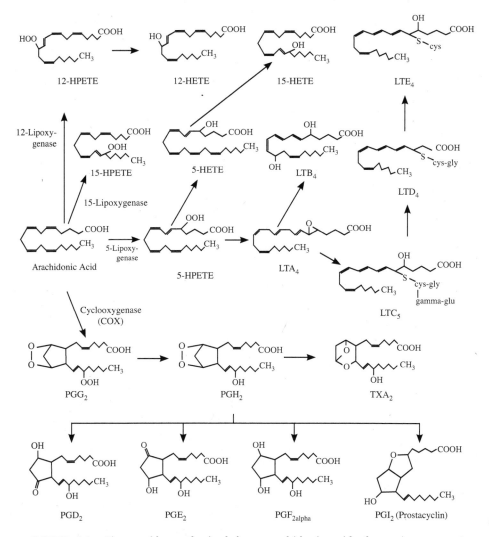

FIGURE C.1. Eicosanoids synthesized from arachidonic acid, the major omega-6 twenty-carbon polyunsaturated fatty acid.

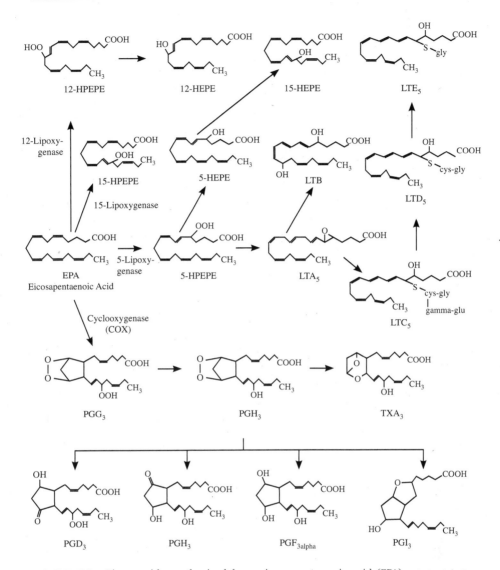

FIGURE C.2. Eicosanoids synthesized from eicosapentaenoic acid (EPA), an omega-3 twenty-carbon polyunsaturated fatty acid.

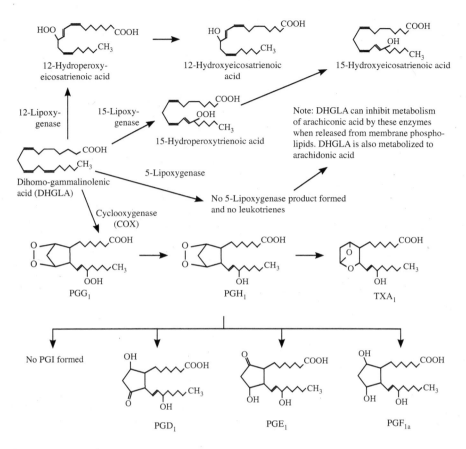

FIGURE C.3. Eicosanoids synthesized from dihomo-gamma-linolenic acid (DHGLA), a minor omega-6 twenty-carbon polyunsaturated fatty acid.

APPENDIX D.
HUMAN LIPOPROTEINS:
THEIR COMPONENTS, PROPERTIES,
AND ASSOCIATED PROTEINS

Lipoproteins transport lipids through the bloodstream. Because the lipids (fats and oils) do not mix with water, they must be packaged into lipoprotein particles that consist of phospholipids and proteins on the surface of the particles, with triglycerides, cholesterol, cholesteryl esters, and other lipids in the core of the particles. The major lipoproteins are chylomicra (or chylomicrons), very low-density lipoprotein (VLDL), low-density lipoprotein (LDL), which is sometimes divided into subpopulations of intermediate-density lipoprotein (IDL or LDL$_1$) and LDL (or LDL$_2$) and high-density lipoprotein (HDL), which is also divided into subpopulations with HDL$_2$ and HDL$_3$ being the major subgroups. These lipoproteins are discussed throughout the book but especially in the context of heart disease. A simple diagram of a phospholipid is shown in figure D.I. Compositions of the lipoproteins are given in table D.I, their range of densities and sizes (diameter) are given in table D.2, and the apolipoproteins, or protein components of lipoproteins, are shown in table D.3.

Lipoproteins are also separated by electrophoresis and classified according to their electrophoretic mobilities relative to α- and β-globulins. The electrophoretic mobilities are based on relative charge on the lipoprotein particles, which is due primarily to the various protein components of each lipoprotein. Lipoproteins within one density category may have several different protein components (as seen in table D.3). Consequently, these lipoproteins are also categorized as α-n, pre-α-n, β-n, and pre-β-n, where n is an integer (usually from I to 3).

The protein components of lipoproteins are designated apolipoproteins (apo- is a prefix used to indicate the protein portion of a proteinaceous molecule that is lacking other components for its complete makeup and ultimate function). There are several apolipoproteins that are components of the various lipoproteins. Table D.3 lists many of the apolipoproteins and the lipoproteins with which they are associated.

Apolipoprotein   Phospholipids

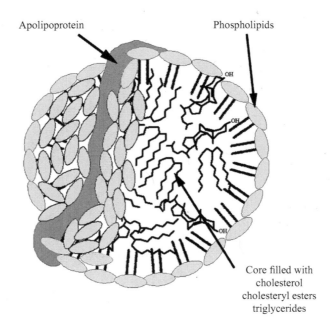

Core filled with
cholesterol
cholesteryl esters
triglycerides

FIGURE D.1.   Schematic illustration of a lipoprotein particle. The particle is surrounded
by phospholipids and apolipoproteins, with cholesterol, cholesteryl esters, triglycerides, and other lipids in its core.

TABLE D.1.
Composition (%) of lipoproteins found in human blood

Lipoprotein	Protein	Phospholipid	Cholesteryl esters	Cholesterol	Triglyceride
Chylomicra	1.5–2.5	7–9	3–5	1–3	84–89
VLDL	5–10	15–20	10–15	5–10	50–65
LDL1 (or IDL)	15–20	22	22	8	30
LDL2 (or LDL)	20–25	15–20	35–40	7–10	7–10
HDL3	50–55	20–25	12	3–4	3
HDL2	40–45	35	12	4	5

Source: Devlin 2002, 129.

TABLE D.2.

Classification of lipoproteins by ultracentrifugation characteristics

Lipoprotein	Density (g/mL)	Diameter (Å)
Chylomicra	< 0.95	10^3 to 10^4
VLDL	0.95–1.006	250–750
LDL1 (or IDL)	1.006–1.019	ca. 250
LDL2 (or LDL)	1.019–1.063	200–280
HDL3	1.063–1.210	50–100
HDL2	1.063–1.210	70–130

Source: Devlin 2002, 127.

TABLE D.3.

Human apolipoproteins and the lipoproteins with which they are associated (values listed are % of total protein in each lipoprotein particle)

Apolipoprotein	HDL2	HDL3	LDL	IDL	VLDL	Chylomicra
ApoA-I	85	70–75	trace	0	0–3	0–3
ApoA-II	5	20	trace	0	0–0.5	0–1.5
Apo B	0–2	—	95–100	50–60	40–50	20–22
ApoC-I	1–2	1–2	0–5	< 1	5	5–10
ApoC-II	1	1	0.5	2.5	10	15
ApoC-III	2–3	2–3	0–5	17	20–25	40
Apo D	0	1–2	—	—	0	1
ApoE	trace	0–5	0	15–20	5–10	5

Source: Devlin 2002, 129.

APPENDIX E.
OVERVIEW OF NEURONAL PROCESSES

There are a few (six or eight) major neurotransmitters that are relatively abundant in the brain. These can be divided into two categories based on chemical properties and physiological functions—amino acids and monoamines. The amino acids, glutamate and GABA (gamma-aminobutyric acid) are by far the most abundant and widely distributed. Glutamate (or glutamic acid) is one of the twenty amino acids used by all cells in the body to make protein. GABA is formed from glutamate by a simple, one-step decarboxylation reaction catalyzed by the enzyme glutamic acid decarboxylase (GAD), as shown in figure E.1. Although these two amino acid transmitters are closely related biochemically, their physiologic actions are diametrically opposed to one another.

Glutamate is the main excitatory neurotransmitter, meaning that it stimulates neurons to fire or release neurotransmitter. GABA is the main inhibitory neurotransmitter, meaning that it restrains neurons and prevents them from firing. The effects of these two transmitters sending their signals to neighboring cells is additive, and a neuron receiving these signals will not fire until it reaches a certain threshold corresponding to numerous incoming excitatory signals. When inhibitory signals predominate, the receiving neuron does not release its neurotransmitter.

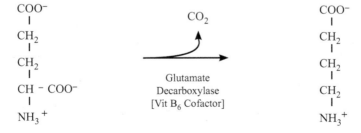

FIGURE E.1. Enzymatic conversion of glutamate to GABA by glutamate decarboxylase (GAD).

228

All cells in the body maintain a certain balance of electrolytes on the inside and outside of their plasma membranes. There are high concentrations of sodium, chloride, and calcium outside of cells and low concentrations of these ions inside a cell, or at least in the cytoplasm of that cell. Calcium can be stored within intracellular compartments and released into the cytoplasm with certain signals. The disparate distribution of ions, particularly sodium (Na+), potassium (K+), calcium (Ca^{2+}), and chloride (Cl$^-$), results in a net charge or voltage across the cell membrane. This voltage is known as membrane potential and functions to carry out vital cell processes, such as bringing nutrients into cells and, in the case of neurons and glandular tissue, sending substances out of a cell. Movement of these ions across membranes of muscle cells regulates contraction and relaxation of a muscle. Movement of electrolytes or ions across neuronal membranes at a nerve ending (presynaptic terminal) controls release of its neurotransmitter.

It was mentioned that glutamate is the main excitatory neurotransmitter in the brain. It can stimulate cells to release transmitter because glutamate receptors on the receiving cells are linked to sodium ion channels (left side of figure E.2). When glutamate binds to its ionotropic receptors, sodium ions are able to rush through the membranes at these sites. This results in loss of charge or voltage across the membrane, and we say that the membrane is depolarized. When the inhibitory transmitter GABA binds to its ionotropic receptors, which are linked to chloride ion channels, chloride is able to rush through the membranes at these sites (right side of figure E.2). Because the inside of the cell

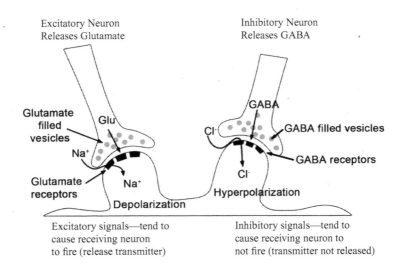

Excitatory Neuron
Releases Glutamate

Inhibitory Neuron
Releases GABA

Glutamate filled vesicles

Glu

Na+

Glutamate receptors

Na+

Depolarization

GABA

Cl-

GABA filled vesicles

GABA receptors

Cl-

Hyperpolarization

Excitatory signals—tend to cause receiving neuron to fire (release transmitter)

Inhibitory signals—tend to cause receiving neuron to not fire (transmitter not released)

FIGURE E.2. Neuronal excitation. Excitatory signals from glutamate cause depolarization by opening sodium channels in the neuronal membranes, and inhibitory signals from GABA cause hyperpolarization by opening chloride channels in the neuronal membranes.

normally has an abundance of negative charges, the additional chloride (negatively charged) ions rushing into the cell result in additional negative charge or greater voltage across the membrane. We say the membrane is hyperpolarized.

At any given time a particular neuron may be receiving signals from numerous other neurons, some of them depolarizing excitatory signals from glutamate-releasing cells and some hyperpolarizing inhibitory signals from GABA-releasing cells. If excitatory signals dominate and the membrane experiences a net depolarization, particularly in the vicinity of the axon connection to the cell body (the axon hillock), the entire axon eventually depolarizes, and the axon terminals release their neurotransmitters. On the other hand, if the inhibitory signals are dominating and the membrane remains polarized or hyperpolarized, there will be no action along the axon, and neurotransmitter is not released. A simple illustration of a neuron is shown in figure E.3.

The incoming excitatory signals conveyed by glutamate are along the dendrites or the cell body. Because the movement of sodium across the cell membrane in response to these signals is dependent on glutamate binding to its receptors, the ion channels linked to the glutamate receptors are known as ligand-gated ion channels. In other words a ligand such as glutamate is needed to open the gates to allow sodium to enter. Depolarization of the axon

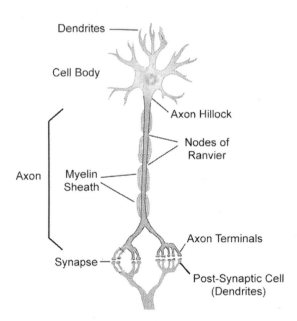

FIGURE E.3.   Basic structure of a neuron, showing the cell body, dendrites, axon, axon hillock, nodes of Ranvier, myelin sheath, and axon terminals. The synapse is a junction (boundary or gap) between an axon terminal of one cell and membrane of another cell. Neurotransmitters emitted by the axon terminal of the presynaptic cell will bind to receptors on the postsynaptic cell (see figure E.5).

does not require glutamate. Once a depolarization threshold is reached at the axon hillock, the rest of the axon becomes depolarized by sodium ions rushing through the axonal membrane at specific sites known as voltage-gated ion channels (see figure E.4). In other words, depolarization (change in voltage) at one site causes voltage-gated channels to open at nearby sites.

Many axons, particularly those of neurons that extend relatively long distances, are covered by a myelin sheath, which acts as an insulator. Voltage-gated sodium channels are located at nodes (sites where the axon is not covered by myelin) that are often separated by relatively long distances—at least by molecular standards. Depolarization by sodium rushing across the axonal membrane at one point causes sodium channels to open beyond the next section of myelin. The entire axonal membrane is not depolarized; depolarization takes place only at the nodes where the axonal membrane is not covered by myelin. Depolarization propagates down a long axon at a rapid rate, which is a hallmark of the nervous system. The electromagnetic field generated by movement of ions across membranes can be detected by sensitive electrodes—the basis for EEGs (electroencephalograms). This same electromagnetic field is responsible for opening sodium ion channels at the next node along the axon.

When depolarization of the axonal membrane reaches the nerve terminals— there may be many terminals on a highly branched axon—there is another type of voltage-gated ion channel that allows calcium ions to enter the nerve terminals (see figure E.5). When calcium rushes into the nerve terminals, it sets

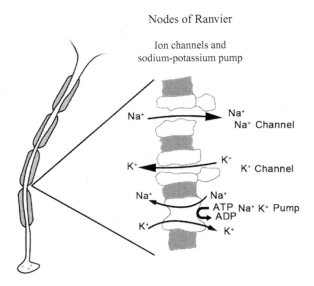

FIGURE E.4. Ions and depolarization. There are high densities of voltage-gated sodium channels and of potassium channels, as well as sodium-potassium pumps (Na/K ATPases), in the axonal membranes at the nodes of Ranvier.

into motion a series of events that ultimately result in release of neurotransmitter from the nerve terminals.

Neurotransmitter molecules are packaged into vesicles (small spheres
surrounded by a membrane). The presence of calcium causes proteins associated with vesicles to fuse with the synaptic membrane and release the vesicle
contents (neurotransmitter) into the synapse. The details of this process are
not essential for discussions in this book, but it is important to understand
that voltage-gated calcium channels may be modified by proteins or enzymes
that have been activated by G-protein-linked (metabotropic) receptors on the
surface of an axon terminal. If voltage-gated calcium channels are modified in a
way that prevents calcium from entering a particular terminal, that terminal will
not release transmitter, even though other terminals along a highly branched
axon are releasing neurotransmitter(s).

There is another way of blocking calcium entry into the nerve terminal
and ultimate release of neurotransmitter. Since the level of potassium is high
inside a cell and low on the outside, there are special types of potassium channels that open in response to certain types of signals. Potassium rushing out of
the cell would cause the membrane to become hyperpolarized. In other words,

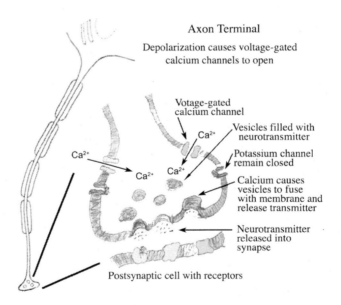

Axon Terminal

Depolarization causes voltage-gated
calcium channels to open

Votage-gated
calcium channel

Vesicles filled with
neurotransmitter

Potassium channel
remain closed

Calcium causes
vesicles to fuse
with membrane and
release transmitter

Neurotransmitter
released into
synapse

Postsynaptic cell with receptors

Figure E.5. Synaptic transmission. When depolarization along the axon (action potential) reaches the nerve terminal, voltage-gated calcium channels are opened, allowing
calcium to rush into the axon terminal. The presence of calcium triggers fusion of
vesicles with the synaptic membrane and release of neurotransmitter. If potassium
channels open in the nerve terminals, that will cancel the depolarization (hyper-
polarize) and prevent the voltage-gated calcium channels from opening.

potassium flowing out of the cell as sodium rushes in would counter the depolarization caused by sodium. These potassium channels are regulated by specific types of transmitter receptors and provide another mechanism for controlling and fine-tuning release of neurotransmitters.

The amine neurotransmitters, such as dopamine, norepinephrine, acetylcholine, and serotonin, are much less abundant than glutamate or GABA. They are considered to be modulatory neurotransmitters, rather than excitatory or inhibitory (although acetylcholine can be excitatory, particularly in the peripheral nervous system). They elicit their effects by binding to receptors linked to G-proteins that ultimately alter metabolism or the activity of other proteins. These neurotransmitters can modify the actions of excitatory and inhibitory neurotransmitters, and their actions usually last for a longer period of time (on the order of minutes) in contrast to the rapid actions (on the order of milliseconds to seconds) of excitatory and inhibitory signals.

Modulatory signals generally involve activation of enzymes (protein kinases) that attach phosphate to other proteins. When phosphate is attached to an ion channel, it may block that channel or it may make that channel more active. Attaching phosphate to some receptor-linked (ligand-gated) ion channels can cause those channels to be withdrawn from the synapse, diminishing response to neurotransmitter in the synapse. Attaching phosphate to some proteins may cause them to move into the nucleus of a cell for expression of genes to make new proteins.

The myriad of events that can take place in neurons in response to certain signals from other neurons is beyond the scope of this book. This has been a cursory overview of how the brain works. It is intended to provide readers with fundamentals for understanding the role that lipids or dietary fats may play in the nervous system. Advances in molecular technologies are allowing scientists to tease out many specific interactions and relate them to specific functions or functional deficits, but the overall integration will continue to pose challenges for generations to come.

GLOSSARY

adipose fat or adipose tissue The distribution of fat storage (adipose) cells is genetically determined, with some individuals concentrating the cells almost exclusively in the abdominal area (especially men), some concentrating them in the hips and thighs (especially women), while others tend to distribute the adipose cells throughout the body (both men and women). There is some question about how much turnover (removal and redepositing) of fats occurs in adipose tissue.

albumin The most abundant protein in blood plasma, albumin can bind and transport many lipids or lipophilic substances. Fatty acids are transported through the blood on albumin, although this is relatively minor in comparison to the fatty acids bound to triglycerides and phospholipids that are transported by lipoproteins.

aldosterone A steroid hormone secreted by the adrenal cortex that regulates sodium levels in the blood by its effect on sodium resorption in the kidney tubules.

alveoli Air sacs in the lungs that can become constricted in response to certain leukotrienes formed from arachidonic acid acting on the smooth-muscle cells around the alveoli.

androgens Steroid hormones responsible for masculine sex characteristics, such as muscle mass and facial hair. These include testosterone and dehydroepiandrosterone.

antioxidants Substances that prevent the oxidation of important biological components. Antioxidants are usually oxidized in the process of protecting other biological components but are then usually recycled or reduced by enzymes to enable repeated cycles of their oxidation. Many natural substances, such as vitamins A, C, and E, as well as phytochemicals (derived from plants) are classified as antioxidants.

apolipoproteins The protein components of lipoproteins are designated apoli-poproteins, where apo- is a prefix used to indicate the protein portion of a pro-teinaceous molecule that is lacking other components for its complete makeup and ultimate function. There are several apolipoproteins that are components of the various lipoproteins. See appendix D for a more complete description of apolipoproteins and lipoproteins.

apoptosis When cells receive damage to critical components, there seem to be signals within the cell for its own demise, or programmed cell death (apopto-sis). This is self-destruction of a cell, which may provide some advantage to the organism as a whole.

arachidonic acid One of the most important fatty acids in the body, in terms of its biochemical and physiological effects, is represented symbolically as C20:4, delta-5,8,11,14. It contains twenty carbons and has four carbon-carbon double bonds. It is classified an omega-6 essential fatty acid and metabolized to numer-ous bioactive eicosanoids and related compounds.

asthma An obstructive pulmonary disease that involves constriction of the deeper airways of the lungs, which is triggered by contraction of smooth mus-cles surrounding airways in response to leukotrienes formed from arachidonic acid.

atherosclerosis The accumulation of modified or oxidized low-density lipopro-tein (oxidized LDL) in the lining of the arteries between endothelial cells and smooth muscle results in fatty deposits known as stenosis or atherosclerosis. This results in reduced volume in the arteries and is usually accompanied by high blood pressure. The fatty deposits may be encapsulated in fibrous proteins to form plaque. Breakdown of the fibrous cap over these fatty deposits exposes the connective proteins to blood platelets, triggering formation of thromboxane A_2 that causes platelet aggregation and initiates a blood clot.

autoimmune disease Autoimmune disease results when the immune system forms antibodies against the body's own tissues or components. This is usually accompanied by inflammation and destruction of the tissue that is causing the immune response.

bile The bile salts, cholesterol, lecithin, and waste products such as bilirubin (from breakdown of heme in hemoglobin) form the bile that is stored in the gallbladder. Bile is usually highly concentrated in bile salts before it is secreted into the small intestine (duodenum) after a meal is consumed. Cholesterol may precipitate or condense in the bile to form gallstones. About 75 percent of gall-stones are cholesterol aggregates; the other 25 percent are from bilirubin. Con-sumption of bile salts may help to break down (dissolve) gallstones.

bile acids/bile salts Cholesterol is oxidized in the liver to form cholic acid, deoxycholic acid, and chenodeoxycholic acid, among others. These are collectively known as bile acids. Like all organic acids, they will ionize in the neutral medium of body fluids, forming the corresponding salts that have the -ate ending (such as cholate and deoxycholate). The bile salts act as detergents in the digestive system, along with phospholipids such as lecithin, transforming large dietary fat globules into an emulsion.

blood-brain barrier The blood-brain barrier results from the tight packing of cells lining the arteries and veins in the brain, requiring all substances that enter the brain or cerebrospinal fluid to pass through these specialized cells rather than seeping through the spaces between cells as in other parts of the body.

bronchoconstriction Narrowing of airways in the lungs due to contraction of smooth muscles surrounding those airways.

bronchodilation Widening of the airways in the lungs due to relaxation of the smooth muscles surrounding those airways.

cannabinoids This refers to all substances that bind to the same receptors in the body that bind tetrahydrocannabinol (THC, the active component in marijuana or cannabis). The endogenous cannabinoids are derived from arachidonic acid (and perhaps other long-chain polyunsaturated fatty acids). These have wide-ranging physiological effects in the body and brain.

cell adhesion molecules (CAMs) CAMs are proteins found on the surface of the plasma membrane of most cells that may act as scaffolding or anchoring sites for other proteins such as connective proteins, which hold cells to one another. CAMs can cause immune cells to stick to certain cells, such as those lining arteries, to initiate the movement of immune cells out of the bloodstream and into specific tissues.

cephalins These are phosphatides composed of glycerol, phosphate, two fatty acids, and ethanolamine esterified to the phosphate group.

cholesterol The mother of all steroids in the body. Cholesterol is incorporated into cell membranes and is metabolized to a wide range of substances, including steroid hormones (progesterone, testosterone, cortisol, cortisone, estrogens), vitamin D (the sunshine vitamin), as well as bile acids or bile salts. Much of the cholesterol in lipoproteins is in the form of cholesteryl esters.

cholesteryl esters Composed of cholesterol combined with a fatty acid, usually an unsaturated fatty acid, forming an ester.

conjugated linoleic acid (CLA) This form of linoleic acid differs from the common fatty acid found in many vegetable oils and other sources in that it contains a trans and a cis double bond rather than two cis double bonds. Conjugated double bonds occur adjacent to one another, which gives them slightly different chemical properties. Conjugated linoleic acids (there are a few isomers) are found in beef, dairy products, and products from other ruminants and have been shown to have anticancer activity.

coronary artery disease (CAD) or coronary heart disease (CHD) CAD is characterized by buildup of atherosclerotic plaque in the arteries supplying the heart muscle, resulting in narrowing of these arteries and decreased oxygen supply that can cause angina pectoris (chest pains) and may eventually result in myocardial infarction (blood clot in the heart).

cyclooxygenase (COX) This enzyme (or group of enzymes) catalyzes reactions of polyunsaturated fatty acids with molecular oxygen to produce prostaglandins G_2 and H_2 from arachidonic acid, which are further metabolized to other prostaglandins and thromboxanes. A wide range of drugs that inhibit this enzyme are collectively known as nonsteroidal anti-inflammatory drugs (NSAIDs).

cytokines Most tissues of the body can produce cytokines (protein messengers) that usually stimulate the immune system. Cytokines are also part of the normal growth and development in the body, providing signals for a wide range of responses. The cytokines include interferons, interleukins, tumor necrosis factor-alpha (TNF-α), as well as many other growth factors and cell adhesion factors.

desaturases The fatty acids are initially synthesized as saturated fatty acids, and then specific desaturase enzymes place double bonds at specific sites in the chain to make them unsaturated. There are at least three different enzymes in humans that place double bonds in positions 5, 6, and 9 of the carbon chain of fatty acids.

dihomo-gamma-linolenic acid This fatty acid is a relatively minor component of dietary fats and oils. It is a metabolic intermediate in the conversion of linoleic acid to arachidonic acid. It can be metabolized to bioactive eicosanoids, but the bioactive products may have different activity than those from arachidonic acid.

docosahexaenoic acid (DHA) This long-chain omega-3 polyunsaturated fatty acid contains twenty-two (*dokosi* in Greek) carbon atoms, with six (-*hexa*-) double bonds (-*en*-). This fatty acid is important in development of the brain and retina in the eye.

eicosanoids Eicosa- is an anglicization of *ikosi*, the Greek word for twenty, and is used in the names of many derivatives of fatty acids that contain twenty carbon atoms. Many of the eicosanoids are potent bioactive molecules. The eicosanoid family includes prostaglandins, thromboxanes, leukotrienes, lipoxins, and cannabinoids, with several members in each of these groups.

eicosapentaenoic acid (EPA) This polyunsaturated fatty acid, along with arachidonic acid, is one of the major precursors for the bioactive eicosanoids. It got its name because it has twenty carbon atoms (*eicosa-*), with five (*-penta-*) double bonds (*-en-*). EPA is a member of the omega-3 fatty acid family, meaning it has double bonds extending to the third carbon from the end of the fatty acid chain farthest from the acid group.

emulsification The suspension of lipids in an aqueous (watery) medium through the detergent action of phospholipids and bile salts.

estrogens Steroid hormones that give rise to female sex characteristics.

extracellular Refers to materials (proteins, fluid, signaling agents) outside the plasma membrane of a cell.

fat A term that refers to a class of lipids known as triglycerides, which are storage fats in living organisms and a major source of calories in our food supply. Traditionally, "fat" referred to triglycerides that are solid at room temperature in contrast to oils, which are liquid at room temperature.

fatty acids Fatty acids are the simplest of lipids and are components of several other classes of lipids, including the triglycerides, phospholipids, glycolipids, cholesteryl esters, and waxes.

fluidity Components of cell membranes, such as receptors, transporters, enzymes, and structural proteins, must be able to move about within the membrane in order to function properly. Lateral movement within the membrane is referred to as fluidity.

foam cells These are macrophages lining the arteries with large deposits of lipid from having engulfed large quantities of oxidized LDL. Their appearance under a microscope is such that they look as if they are filled with foam. These form the fatty streaks that are a prelude to atherosclerosis.

free radicals Highly reactive substances formed by removing or adding an atom or electron from or to the parent substance. Many of the free radicals are derived from oxygen in the atmosphere and are oxidizing substances. Antioxidants prevent the destructive effects of free radicals.

glucocorticoids Steroid hormones produced in the adrenal cortex that influence cellular metabolism and affect the body's response to stress. These include cortisol and cortisone.

G-proteins The G-proteins are generally associated with receptors for hormones, neurotransmitters, and other signaling agents (e.g., eicosanoids). They bind guanosine triphosphate (GTP) when activated and in turn activate other proteins or enzymes inside the cell.

hemostasis Stopping or controlling blood leakage from an injured blood vessel.

homeostasis Maintenance of the chemical and physical conditions of the body and the components of the body fluids, such as electrolytes, proteins, and water, to keep the internal environment of the body suitable for normal life functions.

hormone A chemical mediator that is secreted by a gland and travels throughout the body via the bloodstream to affect or influence specific target cells.

hypoxia A condition of insufficient oxygen, usually caused by a blood clot to a specific organ or tissue. Paradoxically, the hypoxic condition will usually result in oxidative damage to tissue.

immune system A diverse ensemble of mobile and stationary cells in the body that impart the ability to resist or eliminate potentially harmful foreign substances and/or damaged and abnormal cells from within the body. Recognition and destruction of aberrant cells that may be destined to become cancerous tumors is known as immune surveillance.

inflammation Swelling and increased temperature that arises from a series of interrelated events involving immune cells attacking foreign substances or damaged tissue.

intracellular Refers to any substances or activities inside the plasma membrane of a cell.

ischemia Lack of blood supply to a tissue, usually caused by a blood clot or restriction of blood vessels supplying that tissue.

lecithin This phospholipid is composed of glycerol, two fatty acids, phosphate, and choline and is also known as phosphatidyl choline. Lecithin is a component of membranes and lipoproteins, and acts as an emulsifying agent in the bile for digestion of fats.

leukotrienes These comprise one branch of the bioactive eicosanoid family that includes LTB_4, a chemotactic agent for white blood cells (leukocytes), LTC_4, LTD_4, and LTE_4, which are potent bronchoconstrictors, causing difficulty in breathing in an acute allergic or asthmatic attack. Leukotrienes with a 4 subscript are derived from arachidonic acid, while leukotrienes from eicosapentaenoic acid (EPA) have a 5 subscript, reflecting the number of double bonds in the respective leukotrienes.

linoleic acid The most abundant polyunsaturated fatty acid in vegetable oils such as corn, soy, sunflower, and safflower oils. This is an omega-6 essential fatty acid and is metabolized in the body to arachidonic acid, or when consumed in excess may be metabolized for energy. It is highly susceptible to lipid peroxidation.

linolenic acid An omega-3 polyunsaturated fatty acid found in small, varying amounts in some vegetable oils, fish, seafood, and other food sources. It is an essential fatty acid and is metabolized to eicosapentaenoic acid (EPA) and docosahexaenoic acid (DHA) in the body.

lipases Enzymes that catalyze the hydrolysis of fatty acids from triglycerides for mobilization and metabolism of the fatty acids.

lipid An all-encompassing term that includes fats, oils, fatty acids, phospholipids, steroids, fat-soluble vitamins, and other oily substances that dissolve in nonpolar organic solvents, such as hexane or chloroform, but do not dissolve in water.

lipid peroxidation When polyunsaturated fatty acids are exposed to free radicals and oxygen, they react to form a wide range of oxidized products, including fatty acid hydroperoxides (hence the term peroxidation), aldehydes, conjugated dienes, and malondialdehyde. The process of lipid peroxidation can destroy cell membranes and produce oxidized LDL, which is responsible for atherosclerosis.

lipoproteins These transport lipids through the bloodstream. Lipoprotein particles consist of phospholipids and proteins on the surface with triglycerides, cholesterol, cholesteryl esters, and other lipids in the core of the particle. The major lipoproteins are chylomicra (or chylomicrons), very low-density lipoprotein (VLDL), low-density lipoprotein (LDL), and high-density lipoprotein (HDL). The composition and densities for lipoprotein are given in appendix D.

liposomes These are particles composed of a bilayer phospholipid membrane with water on both sides of the bilayer. They have an inner aqueous compartment and are surrounded by water, in contrast to micelles, which have a lipid center. They are used to model cell membranes and have been used to encapsulate certain drugs for delivery to target cells in the body.

lipoxins Lipoxins are a class of bioactive eicosanoids produced from arachidonic acid and eicosapentaenoic acid. They are produced by many different cells and can have a wide range of effects in different tissues.

lipoxygenases These enzymes catalyze reactions between polyunsaturated fatty acids and molecular oxygen to produce a wide range of products, including leukotrienes, lipoxins, resolvins, and neuroprotectins.

macrophages These are immune cells that originate as monocytes in the circulation but become attached to specific tissues and engulf foreign substances or damaged tissue. Macrophages surrounding the arteries, especially the coronary arteries, are responsible for removing oxidized LDL from the circulation. Arterial macrophages with large lipid deposits from engulfed, oxidized LDL are known as foam cells because of their appearance under a microscope.

membrane potential The voltage (potential) resulting from uneven distribution of ions (electrolytes) across a membrane. Cells generally have an abundance of sodium and chloride outside the cell, with potassium and negatively charged proteins inside the cell. This results in a voltage difference of about 60 millivolts across the membrane when the cell is at rest, with the inside of the cell more negative and the outside more positive.

micelles Lipid-filled spheres with shells of phospholipids and perhaps other components such as bile salts. Micelles are distinguished from liposomes in that they have a nonpolar core rather than the bilayer membrane found in liposomes, which have an aqueous core.

monocytes Immune cells in the circulation that, upon receiving certain signals, will migrate to specific tissues where they take up residence as macrophages, engulfing foreign substances or damaged tissue.

myocardial infarction (MI) A blood clot in the arteries that supply the heart muscle. Myocardial infarction can be life-threatening if the clot is not dissolved quickly. This is the medical terminology for a heart attack.

nervous system One of the body's major regulatory systems, the nervous system responds to the external environment by coordinating rapid changes or activities in the body. It consists of the central nervous system (brain) and the peripheral nervous system (the spinal cord, sensory nerves, and other nerves throughout the body).

neuromodulators Chemical messengers that bind to receptors in the nervous system (usually not at the synapse) and cause long-term changes in the response of neurons to incoming signals.

neuropeptides Small proteins that act as hormones and neurotransmitters for signaling in the body. Many of the neuropeptides are released along with classical neurotransmitters at the synapse to effect long-term actions at receptors on nerve cells. Many are also classified as neuromodulators, neurohormones, and neurotransmitters.

nitric oxide (NO) A gas composed of one atom of nitrogen and one atom of oxygen. Although it is a noxious air pollutant, it is produced as a signaling agent in the body from the amino acid arginine. Nitric oxide exerts a wide range of physiological effects in the body, including actions as a neuromodulator and dilation of arteries by causing relaxation of smooth muscles surrounding the arteries.

oil This term generally refers to triglycerides that are in liquid form at room temperature, but there are also many substances known as essential oils that are not necessarily triglycerides.

oleic acid The most abundant monounsaturated fatty acid in nature, with eighteen carbon atoms in the chain and one double bond between C9 and C10. It is an omega-9 fatty acid. Like saturated fatty acids, it is not susceptible to lipid peroxidation, but unlike its saturated counterparts, it imparts some fluidity to membranes and fats when it is present. Olive oil and canola oil are rich sources of this monounsaturated fatty acid. The agricultural industry has developed other vegetable oils with high oleic acid content.

omega-3 and omega-6 Polyunsaturated fatty acids with double bonds extending to the third or sixth carbon, respectively, from the methyl end of the fatty acid chain.

oxidation A chemical reaction that decreases the number of electrons, decreases the number of hydrogen atoms, or increases the amount of oxygen bound to an organic molecule. This type of reaction is characteristic of chemical changes that occur in an atmosphere of oxygen and often involves oxygen, either directly or indirectly.

palmitic acid The most abundant saturated fatty acid found in human tissues and in most food sources, with high levels in palm oil, hence its name. It contains sixteen carbon atoms.

paracrine This term refers to chemical messengers that are secreted by specific cells and have an effect on many cells in a local area, in contrast to hormones, which travel throughout the body in the blood, and neurotransmitters, which have an effect on adjacent cells across a synapse (a narrow space between two nerve cells). Many lipids involved in signaling are paracrine messengers.

peptide Refers to small protein substances composed of several (generally between three and sixty) amino acids linked together in peptide bonds. There are neuropeptides (neurotransmitters) and peptide hormones, with many peptides functioning as both.

phagocytosis The engulfing (endocytosis) of large or complex multimolecular particles by a cell.

phosphatides Phospholipids containing glycerol, two fatty acids, phosphate, and usually an alcohol group that is esterified to the phosphate. When it lacks the alcohol group it is called phosphatidic acid. Phosphatides include lecithin (phosphatidyl choline), phosphatidyl ethanolamine, phosphatidyl serine, and phosphatidyl inositol; the latter usually having additional phosphates on the inositol group.

phospholipids Polar lipids that form bilayer lipid membranes and monolayers surrounding lipoproteins and micelles. The phospholipids include phosphatides, plasmalogens, and sphingomyelins.

phosphorylation This term refers to the process of attaching a phosphate group to a molecule. Phosphorylation is catalyzed by enzymes, which are often activated by signals from molecular messengers. The phosphate is usually derived from adenosine triphosphate (ATP) but may come from other sources. Phosphorylation can affect the activities of enzymes to dramatically alter cellular metabolism, and may affect other proteins such as receptors, ion channels, structural proteins, and transporters.

plasma When blood is collected with an anticoagulant and centrifuged, the red and white blood cells will settle to the bottom of the centrifuge tube. The supernatant, which contains most of the lipoproteins, proteins. and smaller molecules such as sugar, hormones, and electrolytes is known as plasma. It differs from serum in that it still contains fibrinogen and other clotting factors.

plasmalogens These phospholipids contain glycerol, phosphate, one fatty acid, and one alkyl group connected to the glycerol through an ether linkage. The alkyl group often has a double bond near the ether linkage, which gives it antioxidant properties. The high levels of these phospholipids in tissues with high levels of docosahexaenoic acid (DHA) suggest that they are protecting the DHA from lipid peroxidation.

platelets Specialized cells (or cell fragments) in the blood that contain no nucleus or other organelles but have enzymes for certain types of metabolism. They are responsible for initiating a blood clot. The platelets have enzymes to convert arachidonic acid to thromboxane A_2, which causes platelets to aggregate and causes smooth muscles surrounding the local arteries to contract and constrict the blood vessel for clot formation.

polyunsaturated fatty acids Acids that have multiple (two or more) double bonds between carbon atoms in the fatty acid chain. These tend to impart greater fluidity to fats and membrane phospholipids when they are present in abundance.

postsynaptic and presynaptic neurons Neurons form synapses (specialized narrow spaces between cells) where presynaptic neurons release neurotransmitters that migrate across the synapse to bind receptors on the postsynaptic cells (neurons may communicate with many types of receptive cells, such as neurons, glands, and muscles).

prostaglandins A branch of the bioactive eicosanoids that include PGD_2, PGE_2, and $PGF_{2\alpha}$. The 2 subscript indicates there are two carbon-carbon double bonds in the structure, whereas a 3 subscript would represent three double bonds and a 1 subscript would indicate one double bond. The 1-series of prostaglandins is derived from dihomo-gamma-linolenic acid (DHGLA), the 2-series is formed from arachidonic acid, and the 3-series from eicosapentaenoic acid (EPA). The different series of prostaglandins will display similar physiological actions, but with subtle differences in activity.

protein kinases Enzymes that catalyze the addition of phosphate (phosphorylation) to specific proteins. ATP is the source of the phosphate group that is transferred to the target protein, and the phosphorylation usually results in a drastic change in the activity or function of the target protein.

proteolytic enzymes or proteases Enzymes that hydrolyze (break down or digest) proteins. Some types of proteases require a metal cofactor for activity and are known as metalloproteases. These are often found in the extracellular fluid where the immune system is active and may damage or begin destroying structural proteins in atherosclerotic plaque to trigger formation of a blood clot. Their normal function is to break down protein when repairing tissue.

receptors Proteins that bind to specific chemicals or messengers and are linked to other proteins that produce some response in the cell where the receptor is located. Receptors may be linked to ion channels that open when the messenger binds the receptor, or they may be linked to other proteins, such as enzymes that catalyze some transformation. The net effect is that something will be altered in the cell as a result of messengers binding to receptors.

saturated fatty acids These acids have no double bonds between carbon atoms in the fatty acid chain. They are not susceptible to lipid peroxidation and are the most common fatty acids in cell membranes, with palmitic acid predominating in the human body.

second messenger An intracellular chemical or ion (e.g., Ca^{2+}) that is produced when a primary messenger (hormone, neurotransmitter, paracrine factor) binds to a receptor that triggers the production of the secondary messenger inside the cell. The second messenger will generally alter the protein by activating or inhibiting its function.

serum When blood is collected and allowed to coagulate, removing large amounts of protein clotting factors, and subsequently centrifuged to remove blood cells, the remaining serum (supernatant) contains lipoproteins, nonclotting proteins, and smaller molecules and electrolytes.

signal transduction This is the net effect of events following the binding of a primary messenger to a receptor, which leads to some change in the cell receiving that signal. It is usually a change in concentration of ions (when ion channels are opened) or in the activities of proteins when second messengers are produced.

sphingolipids Polar lipids containing sphingosine, a fatty acid, and either a phosphate or carbohydrate group. An alcohol, such as choline, is usually esterified to the phosphate to form sphingomyelins. Cerebrosides have glucose or galactose as the carbohydrate group, and when there are multiple sugars to form an oligosaccharide, the substance is known as a ganglioside.

stearic acid A saturated fatty acid with eighteen carbon atoms in the fatty acid chain. It is abundant in cocoa butter and chocolate but is also found in significant amounts in most animal fats.

steroids Organic chemicals with a characteristic structure containing three hexagonal rings and one pentagonal ring. Cholesterol is the parent steroid in animals. Plant steroids have the same rings in their structures but differ in the groups attached to these rings. Many steroid hormones are formed in the body from cholesterol.

synapse A specialized junction where communication can take place between two cells. The communication is usually in the form of chemicals (neurotransmitters) being released by one (presynaptic) cell and binding to the other (postsynaptic) cell. There are numerous specialized proteins in the membranes of each cell where the synapse is formed.

target cells Extracellular signals (hormones, neurotransmitters, paracrine factors) affect only cells that have receptors for those messengers. If a cell has a receptor for a particular messenger, it is considered a target cell for that messenger.

thromboxanes These are eicosanoids derived from twenty-carbon polyunsaturated fatty acids that cause platelets to aggregate and cause blood vessels to constrict. This sets up further reactions in the cascade that eventually produces a blood clot. The most potent is thromboxane A_2 (TXA_2), which is derived from arachidonic acid, whereas TXA_3, derived from eicosapentaenoic acid, is less potent at platelet aggregation and vasoconstriction.

thrombus A blood clot attached to the inner lining of a blood vessel.

trans fats These are typically monounsaturated fatty acids with trans double bonds. They are found in partially hydrogenated vegetable oils used in margarine and shortening. Trace amounts of trans-fatty acids are found in meat and milk products from ruminants. These fatty acids can increase serum total cholesterol and LDL cholesterol, although their effects (positive or negative) on health are not well established. The food industry has drastically reduced the amount of trans fats in its products.

triglycerides or triacylglycerols Storage lipids containing three (*tri-*) fatty acids (*-acyl-*) esterified to a glycerol molecule. Triglycerides are the major components of dietary fats and oils, and the fats in adipose tissue. They are transported through the blood primarily in the larger lipoprotein particles, chylomicra and VLDL.

unsaturated fatty acids Fatty acids with at least one (monounsaturated) double bond between carbon atoms, as in oleic acid, or with two or more (polyunsaturated) double bonds between carbon atoms in the fatty acid chain, as in linoleic and linolenic acids and their longer-chain derivatives.

vasoconstriction and vasodilation The narrowing and widening of blood vessels in response to specific signals or messengers. Thromboxane A_2 causes constriction of blood vessels by signaling smooth muscles to contract. Prostaglandin I_2 and nitric oxide cause vasodilation by signaling the smooth muscles to relax.

REFERENCES

Abedin, M., J. Lim, T. B. Tang, D. Park, L. L. Demer, and Y. Tintut. 2006. N-3 fatty acids inhibit vascular calcification via the p38-mitogen-activated protein kinase and peroxisome proliferator-activated receptor-gamma pathways. *Circ Res* 98 (6): 727–729.

Abedin, M., Y. Tintut, and L. L. Demer. 2004. Vascular calcification: Mechanisms and clinical ramifications. *Arterioscler Thromb Vasc Biol* 24 (7): 1161–1170.

Adams, J. M., and S. Cory. 1991. Transgenic models of tumor development. *Science* 254 (5035): 1161–1167.

Al, M. D., A. C. van Houwelingen, and G. Hornstra. 1997. Relation between birth order and the maternal and neonatal docosahexaenoic acid status. *Eur J Clin Nutr* 51 (8): 548–553.

Albanes, D., O. P. Heinonen, P. R. Taylor, J. Virtamo, B. K. Edwards, M. Rautalahti, A. M. Hartman, et al. 1996. Alpha-tocopherol and beta-carotene supplements and lung cancer incidence in the alpha-tocopherol, beta-carotene cancer prevention study: Effects of base-line characteristics and study compliance. *J Natl Cancer Inst* 88 (21): 1560–1570.

Albert, M. A., E. Danielson, N. Rifai, and P. M. Ridker. 2001. Effect of statin therapy on C-reactive protein levels: The pravastatin inflammation/CRP evaluation (PRINCE): A randomized trial and cohort study. *JAMA* 286 (1): 64–70.

Alboni, P., E. Favaron, N. Paparella, M. Sciammarella, and M. Pedaci. 2008. Is there an association between depression and cardiovascular mortality or sudden death? *J Cardiovasc Med* (Hagerstown) 9 (4): 356–362.

Ames, B. N., and L. S. Gold. 1991. Carcinogenesis mechanisms: The debate continues. *Science* 252:902.

Antalis, C. J., L. J. Stevens, M. Campbell, R. Pazdro, K. Ericson, and J. R. Burgess. 2006. Omega-3 fatty acid status in attention-deficit/hyperactivity disorder. *Prostaglandins Leukot Essent Fatty Acids* 75 (4–5): 299–308.

Apostolski, S., Z. Marinkovic, A. Nikolic, D. Blagojevic, M. B. Spasic, and A. M. Michelson. 1998. Glutathione peroxidase in amyotrophic lateral sclerosis: The effects of selenium supplementation. *J Environ Pathol Toxicol Oncol* 17 (3–4): 325–329.

Appleton, K. M., R. C. Hayward, D. Gunnell, T. J. Peters, P. J. Rogers, D. Kessler, and A. R. Ness. 2006. Effects of n-3 long-chain polyunsaturated fatty acids on depressed mood: Systematic review of published trials. *Am J Clin Nutr* 84 (6): 1308–1316.

Asztalos, B. F., L. A. Cupples, S. Demissie, K. V. Horvath, C. E. Cox, M. C. Batista, and E. J. Schaefer. 2004. High-density lipoprotein subpopulation profile and coronary heart disease prevalence in male participants of the Framingham Offspring Study. *Arterioscler Thromb Vasc Biol* 24 (11): 2181–2187.

Asztalos, B., M. Lefevre, L. Wong, T. A. Foster, R. Tulley, M. Windhauser, W. Zhang, and P. S. Roheim. 2000. Differential response to low-fat diet between low and normal HDL-cholesterol subjects. *J Lipid Res* 41 (3): 321–328.

Atkins, Robert C. 1972. *Dr. Atkins' diet revolution.* New York: Bantam Books.

Barter, P. J., S. Nicholls, K. A. Rye, G. M. Anantharamaiah, M. Navab, and A. M. Fogelman. 2004. Antiinflammatory properties of HDL. *Circ Res* 95 (8): 764–772.

Barzilai, N., G. Atzmon, C. Schechter, E. J. Schaefer, A. L. Cupples, R. Lipton, S. Cheng, and A. R. Shuldiner. 2003. Unique lipoprotein phenotype and genotype associated with exceptional longevity. *JAMA* 290 (15): 2030–2040.

Basciano, H., L. Federico, and K. Adeli. 2005. Fructose, insulin resistance, and metabolic dyslipidemia. *Nutr Metab* (Lond) 2 (1): 5.

Belury, M. A. 2002. Inhibition of carcinogenesis by conjugated linoleic acid: Potential mechanisms of action. *J Nutr* 132 (10): 2995–2998.

Berger, G. E., S. Smesny, and G. P. Amminger. 2006. Bioactive lipids in schizophrenia. *Int Rev Psychiatry* 18 (2): 85–98.

Berkow, R., ed. 1982. *The Merck manual of diagnosis and therapy.* 14th ed. Rahway, NJ: Merck, Sharp and Dohme Research Laboratories.

Berry, E. M. 1997. Dietary fatty acids in the management of diabetes mellitus. *Am J Clin Nutr* 66 (4 Suppl): 991S–997S.

Bilici, M., F. Yildirim, S. Kandil, M. Bekaroglu, S. Yildirmis, O. Deger, M. Ulgen, A. Yildiran, and H. Aksu. 2004. Double-blind, placebo-controlled study of zinc sulfate in the treatment of attention deficit hyperactivity disorder. *Prog Neuropsychopharmacol Biol Psychiatry* 28 (1): 181–190.

Black, P. N., and S. Sharpe. 1997. Dietary fat and asthma: Is there a connection? *Eur Respir J* 10 (1): 6–12.

Bombardier, C., L. Laine, A. Reicin, D. Shapiro, R. Burgos-Vargas, B. Davis, R. Day, et al. 2000. Comparison of upper gastrointestinal toxicity of rofecoxib and naproxen in patients with rheumatoid arthritis. VIGOR Study Group. *N Engl J Med* 343 (21): 1520–1528.

Booth, F. W., M. J. Laye, S. J. Lees, R. S. Rector, and J. P. Thyfault. 2008. Reduced physical activity and risk of chronic disease: The biology behind the consequences. *Eur J Appl Physiol* 102 (4): 381–390.

Bordoni, A., M. Di Nunzio, F. Danesi, and P. L. Biagi. 2006. Polyunsaturated fatty acids: From diet to binding to ppars and other nuclear receptors. *Genes Nutr* 1 (2): 95–106.

Bough, K. J., and J. M. Rho. 2007. Anticonvulsant mechanisms of the ketogenic diet. *Epilepsia* 48 (1): 43–58.

Bourre, J. M. 2004. Roles of unsaturated fatty acids (especially omega-3 fatty acids) in the brain at various ages and during ageing. *J Nutr Health Aging* 8 (3): 163–174.

Braden, L. M., and K. K. Carroll. 1986. Dietary polyunsaturated fat in relation to mammary carcinogenesis in rats. *Lipids* 21 (4): 285–288.

Brehm, B. J., R. J. Seeley, S. R. Daniels, and D. A. D'Alessio. 2003. A randomized trial comparing a very low carbohydrate diet and a calorie-restricted low fat diet on body weight and cardiovascular risk factors in healthy women. *J Clin Endocrinol Metab* 88 (4): 1617–1623.

Briscoe, C. P., M. Tadayyon, J. L. Andrews, W. G. Benson, J. K. Chambers, M. M. Eilert, C. Ellis, et al. 2003. The orphan G protein-coupled receptor GPR40 is activated by medium and long chain fatty acids. *J Biol Chem* 278 (13): 11303–11311.

Bronte-Stewart, B., A. Antonis, L. Eales, and J. F. Brock. 1956. Effects of feeding different fats on serum-cholesterol level. *Lancet* 270 (6922): 521–526.

Brophy, M. H. 1986. Zinc and childhood hyperactivity. *Biol Psychiatry* 21 (7): 704–705.

Broughton, K. S., C. S. Johnson, B. K. Pace, M. Liebman, and K. M. Kleppinger. 1997. Reduced asthma symptoms with n-3 fatty acid ingestion are related to 5-series leukotriene production. *Am J Clin Nutr* 65 (4): 1011–1017.

Brousseau, M. E., G. P. Eberhart, J. Dupuis, B. F. Asztalos, A. L. Goldkamp, E. J. Schaefer, and M. W. Freeman. 2000. Cellular cholesterol efflux in heterozygotes for tangier disease is markedly reduced and correlates with high density lipoprotein cholesterol concentration and particle size. *J Lipid Res* 41 (7): 1125–1135.

Brown, A. G., T. C. Smale, T. J. King, R. Hasenkamp, and R. H. Thompson. 1976. Crystal and molecular structure of compactin, a new antifungal metabolite from Penicillium brevicompactum. *J Chem Soc* [Perkin I] (11): 1165–1170.

Brown, A. J., S. Jupe, and C. P. Briscoe. 2005. A family of fatty acid binding receptors. *DNA Cell Biol* 24 (1): 54–61.

Brown, M. S., and J. L. Goldstein. 1986. A receptor-mediated pathway for cholesterol homeostasis. *Science* 232 (4746): 34–47.

———. 1997. The SREBP pathway: Regulation of cholesterol metabolism by proteolysis of a membrane-bound transcription factor. *Cell* 89 (3): 331–340.

Brownell, K., and K. B. Horgen. 2004. *Food fight: The inside story of the food industry, America's obesity crisis, and what we can do about it.* New York: McGraw-Hill Companies.

Brull, D. J., N. Serrano, F. Zito, L. Jones, H. E. Montgomery, A. Rumley, P. Sharma, et al. 2003. Human CRP gene polymorphism influences CRP levels: Implications for the prediction and pathogenesis of coronary heart disease. *Arterioscler Thromb Vasc Biol* 23 (11): 2063–2069.

Burgess, J. R., L. Stevens, W. Zhang, and L. Peck. 2000. Long-chain polyunsaturated fatty acids in children with attention-deficit hyperactivity disorder. *Am J Clin Nutr* 71 (1 Suppl): 327S–330S.

Burr, G. O., and M. M. Burr. 1929. A new deficiency disease produced by rigid exclusion of fat from the diet. *J Biol Chem* 82:345–367.

———. 1930. On the nature and role of the fatty acids essential in nutrition. *J Biol Chem* 86:587–621.

Busserolles, J., E. Gueux, E. Rock, A. Mazur, and Y. Rayssiguier. 2002. Substituting honey for refined carbohydrates protects rats from hypertriglyceridemic and prooxidative effects of fructose. *J Nutr* 132 (11): 3379–3382.

Calderon, F., and H. Y. Kim. 2004. Docosahexaenoic acid promotes neurite growth in hippocampal neurons. *J Neurochem* 90 (4): 979–988.

Callister, T. Q., P. Raggi, B. Cooil, N. J. Lippolis, and D. J. Russo. 1998. Effect of HMG-CoA reductase inhibitors on coronary artery disease as assessed by electron-beam computed tomography. *N Engl J Med* 339 (27): 1972–1978.

Campos, H., J. J. Genest, Jr., E. Blijlevens, J. R. McNamara, J. L. Jenner, J. M. Ordovas, P. W. Wilson, and E. J. Schaefer. 1992. Low density lipoprotein particle size and coronary artery disease. *Arterioscler Thromb* 12 (2): 187–195.

Carlson, S. E., S. H. Werkman, J. M. Peeples, R. J. Cooke, and E. A. Tolley. 1993. Arachidonic acid status correlates with first year growth in preterm infants. *Proc Natl Acad Sci USA* 90 (3): 1073–1077.

Carlson, S. E., S. H. Werkman, and E. A. Tolley. 1996. Effect of long-chain n-3 fatty acid supplementation on visual acuity and growth of preterm infants with and without bronchopulmonary dysplasia. *Am J Clin Nutr* 63 (5): 687–697.

Carroll, M. D., D. A. Lacher, P. D. Sorlie, J. I. Cleeman, D. J. Gordon, M. Wolz, S. M. Grundy, and C. L. Johnson. 2005. Trends in serum lipids and lipoproteins of adults, 1960–2002. *JAMA* 294 (14): 1773–1781.

Casset, A., C. Marchand, A. Purohit, S. le Calve, B. Uring-Lambert, C. Donnay, P. Meyer, and F. de Blay. 2006. Inhaled formaldehyde exposure: Effect on bronchial response to mite allergen in sensitized asthma patients. *Allergy* 61 (11): 1344–1350.

Chen, C., and N. G. Bazan. 2005. Lipid signaling: Sleep, synaptic plasticity, and neuroprotection. *Prostaglandins Other Lipid Mediat* 77 (1–4): 65–76.

Cheung, B. M., I. J. Lauder, C. P. Lau, and C. R. Kumana. 2004. Meta-analysis of large randomized controlled trials to evaluate the impact of statins on cardiovascular outcomes. *Br J Clin Pharmacol* 57 (5): 640–651.

Chin, S. F., W. Liu, J. M. Storkson, Y. L. Ha, and M. W. Pariza. 1992. Dietary sources of conjugated dienoic isomers of linoleic acid, a newly recognized class of anticarcinogens. *J Food Compos Anal* 5:185–197.

Chisolm, G. M., G. Ma, K. C. Irwin, L. L. Martin, K. G. Gunderson, L. F. Linberg, D. W. Morel, and P. E. DiCorleto. 1994. 7 beta-hydroperoxycholest-5-en-3 beta-ol, a component of human atherosclerotic lesions, is the primary cytotoxin of oxidized human low density lipoprotein. *Proc Natl Acad Sci USA* 91 (24): 11452–11456.

Choudhury, N., L. Tan, and A. S. Truswell. 1995. Comparison of palmolein and olive oil: Effects on plasma lipids and vitamin E in young adults. *Am J Clin Nutr* 61 (5): 1043–1051.

Christensen, O., and E. Christensen. 1988. Fat consumption and schizophrenia. *Acta Psychiatr Scand* 78 (5): 587–591.

Cimino, P. J., C. D. Keene, R. M. Breyer, K. S. Montine, and T. J. Montine. 2008. Therapeutic targets in prostaglandin E2 signaling for neurologic disease. *Curr Med Chem* 15 (19): 1863–1869.

Clària, J., and M. Romano. 2005. Pharmacological intervention of cyclooxygenase-2 and 5-lipoxygenase pathways: Impact on inflammation and cancer. *Curr Pharm Des* 11 (26): 3431–3447.

Clària, J., and C. N. Serhan. 1995. Aspirin triggers previously undescribed bioactive eicosanoids by human endothelial cell-leukocyte interactions. *Proc Natl Acad Sci USA* 92 (21): 9475–9479.

Cleland, L. G., and M. J. James. 1997. Rheumatoid arthritis and the balance of dietary N-6 and N-3 essential fatty acids. *Br J Rheumatol* 36 (5): 513–514.

Cohen, P., and J. M. Friedman. 2004. Leptin and the control of metabolism: Role for stearoyl-CoA desaturase-1 (SCD-1). *J Nutr* 134 (9): 2455S–2463S.

Cohen, P., C. Zhao, X. Cai, J. M. Montez, S. C. Rohani, P. Feinstein, P. Mombaerts, and J. M. Friedman. 2001. Selective deletion of leptin receptor in neurons leads to obesity. *J Clin Invest* 108 (8): 1113–1121.

Coleman, D. L. 1978. Obese and diabetes: Two mutant genes causing diabetes-obesity syndromes in mice. *Diabetologia* 14 (3): 141–148.

Colquhoun, I., and S. Bunday. 1981. A lack of essential fatty acids as a possible cause of hyperactivity in children. *Med Hypotheses* 7 (5): 673–679.

Commentary. 2000. Inflammation and Alzheimer's disease: Open peer commentary. *Neurobiol Aging* 21:383–421.

Consilvio, C., A. M. Vincent, and E. L. Feldman. 2004. Neuroinflammation, COX-2, and ALS—a dual role? *Exp Neurol* 187 (1): 1–10.

Cordain, Loren. 2009. Table of fatty acids in vegetable oils. The Paleo Diet 2009 [accessed June 30, 2009]. http://www.thepaleodiet.com/nutritional_tools/oils_table.html.

Corsini, A. 2003. The safety of HMG-CoA reductase inhibitors in special populations at high cardiovascular risk. *Cardiovasc Drugs Ther* 17 (3): 265–285.

Cotzias, G. C. 1971. Levodopa in the treatment of Parkinsonism. *JAMA* 218 (13): 1903–1908.

Cowburn, A. S., K. Sladek, J. Soja, L. Adamek, E. Nizankowska, A. Szczeklik, B. K. Lam, et al.

1998. Overexpression of leukotriene C4 synthase in bronchial biopsies from patients with aspirin-intolerant asthma. *J Clin Invest* 101 (4): 834–846.

Cravatt, B. F., O. Prospero-Garcia, G. Siuzdak, N. B. Gilula, S. J. Henriksen, D. L. Boger, and R. A. Lerner. 1995. Chemical characterization of a family of brain lipids that induce sleep. *Science* 268 (5216): 1506–1509.

Cynshi, O., and R. Stocker. 2005. Inhibition of lipoprotein lipid oxidation. *Handb Exp Pharmacol* (170):563–590.

Daly, M. 2003. Sugars, insulin sensitivity, and the postprandial state. *Am J Clin Nutr* 78 (4): 865S–872S.

D'Angelo, A., and J. Selhub. 1997. Homocysteine and thrombotic disease. *Blood* 90 (1): 1–11.

Davidson, M. H., and J. G. Robinson. 2007. Safety of aggressive lipid management. *J Am Coll Cardiol* 49 (17): 1753–1762.

Davignon, J., and L. A. Leiter. 2005. Ongoing clinical trials of the pleiotropic effects of statins. *Vasc Health Risk Manag* 1 (1): 29–40.

de la Presa Owens, S., and S. M. Innis. 1999. Docosahexaenoic and arachidonic acid prevent a decrease in dopaminergic and serotoninergic neurotransmitters in frontal cortex caused by a linoleic and alpha-linolenic acid deficient diet in formula-fed piglets. *J Nutr* 129 (11): 2088–2093.

———. 2000. Diverse, region-specific effects of addition of arachidonic and docosahexanoic acids to formula with low or adequate linoleic and alpha-linolenic acids on piglet brain monoaminergic neurotransmitters. *Pediatr Res* 48 (1): 125–130.

Demer, L. L., and Y. Tintut. 2003. Mineral exploration: Search for the mechanism of vascular calcification and beyond: The 2003 Jeffrey M. Hoeg Award lecture. *Arterioscler Thromb Vasc Biol* 23 (10): 1739–1743.

Denke, M. A., and S. M. Grundy. 1992. Comparison of effects of lauric acid and palmitic acid on plasma lipids and lipoproteins. *Am J Clin Nutr* 56 (5): 895–898.

Deutsch, D. G., M. S. Goligorsky, P. C. Schmid, R. J. Krebsbach, H. H. Schmid, S. K. Das, S. K. Dey, et al. 1997. Production and physiological actions of anandamide in the vasculature of the rat kidney. *J Clin Invest* 100 (6): 1538–1546.

Devane, W. A., L. Hanus, A. Breuer, R. G. Pertwee, L. A. Stevenson, G. Griffin, D. Gibson, A. Mandelbaum, A. Etinger, and R. Mechoulam. 1992. Isolation and structure of a brain constituent that binds to the cannabinoid receptor. *Science* 258 (5090): 1946–1949.

Devlin, Thomas M. 2002. *Textbook of biochemistry with clinical correlations.* New York: Wiley-Liss.

Drachman, D. B., K. Frank, M. Dykes-Hoberg, P. Teismann, G. Almer, S. Przedborski, and J. D. Rothstein. 2002. Cyclooxygenase 2 inhibition protects motor neurons and prolongs survival in a transgenic mouse model of ALS. *Ann Neurol* 52 (6): 771–778.

Dupuis, L., J. L. Gonzalez de Aguilar, H. Oudart, M. de Tapia, L. Barbeito, and J. P. Loeffler. 2004. Mitochondria in amyotrophic lateral sclerosis: A trigger and a target. *Neurodegener Dis* 1 (6): 245–254.

Dyerberg, J., and H. O. Bang. 1980. All cis-5, 8, 11, 14, 17 eicosapentaenoic acid and triene prostaglandins: Potential anti-thrombotic agents. *Scand J Clin Lab Invest* 40 (7): 589–593.

Eichner, J. E., S. T. Dunn, G. Perveen, D. M. Thompson, K. E. Stewart, and B. C. Stroehla. 2002. Apolipoprotein E polymorphism and cardiovascular disease: A HuGE review. *Am J Epidemiol* 155 (6): 487–495.

Elliott, S. S., N. L. Keim, J. S. Stern, K. Teff, and P. J. Havel. 2002. Fructose, weight gain, and the insulin resistance syndrome. *Am J Clin Nutr* 76 (5): 911–922.

el-Mesallamy, H., S. Suwailem, and N. Hamdy. 2007. Evaluation of C-reactive protein, endothelin-1, adhesion molecule(s), and lipids as inflammatory markers in type 2 diabetes mellitus patients. *Mediators Inflamm* 2007, article no. 73635.

Emerit, J., M. Edeas, and F. Bricaire. 2004. Neurodegenerative diseases and oxidative stress. *Biomed Pharmacother* 58 (1): 39–46.

Endo, A. 1992. The discovery and development of HMG-CoA reductase inhibitors. *J Lipid Res* 33 (11): 1569–1582.

Endres, S., R. Lorenz, and K. Loeschke. 1999. Lipid treatment of inflammatory bowel disease. *Curr Opin Clin Nutr Metab Care* 2 (2): 117–120.

Erickson, K. L. 1998. Is there a relation between dietary linoleic acid and cancer of the breast, colon, or prostate? *Am J Clin Nutr* 68 (1): 5–7.

Ferris, C. F., S. F. Lu, T. Messenger, C. D. Guillon, N. Heindel, M. Miller, G. Koppel, F. Robert Bruns, and N. G. Simon. 2006. Orally active vasopressin V1a receptor antagonist, SRX251, selectively blocks aggressive behavior. *Pharmacol Biochem Behav* 83 (2): 169–174.

Fleming, D. M., and D. L. Crombie. 1987. Prevalence of asthma and hay fever in England and Wales. *Br Med J (Clin Res Ed)* 294 (6567): 279–283.

Fride, E. 2002. Endocannabinoids in the central nervous system—an overview. *Prostaglandins Leukot Essent Fatty Acids* 66 (2–3): 221–233.

Fried, S. K., and S. P. Rao. 2003. Sugars, hypertriglyceridemia, and cardiovascular disease. *Am J Clin Nutr* 78 (4): 873S–880S.

Gaoni, Y., and R. Mechoulam. 1964. Isolation, structure, and partial synthesis of an active constituent of hashish. *J Amer Chem Soc* 86:1646.

Garcia Rodriguez, L. A., and A. Gonzalez-Perez. 2004. Inverse association between nonsteroidal anti-inflammatory drugs and prostate cancer. *Cancer Epidemiol Biomarkers Prev* 13 (4): 649–653.

Garcia-Rodriguez, L. A., and C. Huerta-Alvarez. 2001. Reduced risk of colorectal cancer among long-term users of aspirin and nonaspirin nonsteroidal antiinflammatory drugs. *Epidemiology* 12 (1): 88–93.

Gardner, C. D., A. Kiazand, S. Alhassan, S. Kim, R. S. Stafford, R. R. Balise, H. C. Kraemer, and A. C. King. 2007. Comparison of the Atkins, Zone, Ornish, and LEARN diets for change in weight and related risk factors among overweight premenopausal women: The A TO Z Weight Loss Study: A randomized trial. *JAMA* 297 (9): 969–977.

Garlid, K. D., M. Jaburek, and P. Jezek. 2001. Mechanism of uncoupling protein action. *Biochem Soc Trans* 29 (6): 803–806.

Geyelin, H. R. 1921. Fasting as a method of treating epilepsy. *Medical Record* 99:1037–1039.

Glew, Robert H. 2002. Lipid metabolism II: Pathways of metabolism of special lipids. In *Textbook of biochemistry with clinical correlations*. New York: Wiley-Liss.

Goldner, E. M., L. Hsu, P. Waraich, and J. M. Somers. 2002. Prevalence and incidence studies of schizophrenic disorders: A systematic review of the literature. *Can J Psychiatry* 47 (9): 833–843.

Goldstein, J. L., and M. S. Brown. 2001. Molecular medicine. The cholesterol quartet. *Science* 292 (5520): 1310–1312.

Gould, A. L., G. M. Davies, E. Alemao, D. D. Yin, and J. R. Cook. 2007. Cholesterol reduction yields clinical benefits: Meta-analysis including recent trials. *Clin Ther* 29 (5): 778–794.

Griffin, B. A., D. J. Freeman, G. W. Tait, J. Thomson, M. J. Caslake, C. J. Packard, and J. Shepherd. 1994. Role of plasma triglyceride in the regulation of plasma low density lipoprotein (LDL) subfractions: Relative contribution of small, dense LDL to coronary heart disease risk. *Atherosclerosis* 106 (2): 241–253.

Grundy, S. M., J. I. Cleeman, C. N. Merz, H. B. Brewer, Jr., L. T. Clark, D. B. Hunninghake, R. C.

Pasternak, S. C. Smith, Jr., and N. J. Stone. 2004. Implications of recent clinical trials for the National Cholesterol Education Program Adult Treatment Panel III guidelines. *Circulation* 110 (2): 227–239.

Ha, Y. L., N. K. Grimm, and M. W. Pariza. 1987. Anticarcinogens from fried ground beef: Heat-altered derivatives of linoleic acid. *Carcinogenesis* 8 (12): 1881–1887.

Ha, Y. L., J. Storkson, and M. W. Pariza. 1990. Inhibition of benzo(a)pyrene-induced mouse forestomach neoplasia by conjugated dienoic derivatives of linoleic acid. *Cancer Res* 50 (4): 1097–1101.

Hahnel, D., J. Thiery, T. Brosche, and B. Engelmann. 1999. Role of plasmalogens in the enhanced resistance of LDL to copper-induced oxidation after LDL apheresis. *Arterioscler Thromb Vasc Biol* 19 (10): 2431–2438.

Haley, S., J. Reed, B.-H. Lin, and A. Cook. 2005. Sweetener consumption in the United States: Distribution by demographics and product characteristics. U.S. Department of Agriculture 2005 [accessed June 30, 2009]. http://www.ers.usda.gov/Publications/SSS/aug05/sss24301/sss24301.pdf.

Hallahan, B., and M. R. Garland. 2004. Essential fatty acids and their role in the treatment of impulsivity disorders. *Prostaglandins Leukot Essent Fatty Acids* 71 (4): 211–216.

Harris, R. E., J. Beebe-Donk, and G. A. Alshafie. 2006. Reduction in the risk of human breast cancer by selective cyclooxygenase-2 (COX-2) inhibitors. *BMC Cancer* 6:27.

Hartl, D. L., and E. W. Jones. 1991. *Genetics: Analysis of genes and genomes.* 5th ed. Sudbury, MA: Jones and Bartlett Publishers.

Hayes, C. E., M. T. Cantorna, and H. F. DeLuca. 1997. Vitamin D and multiple sclerosis. *Proc Soc Exp Biol Med* 216 (1): 21–27.

Hegsted, D. M., R. B. McGandy, M. L. Myers, and F. J. Stare. 1965. Quantitative effects of dietary fat on serum cholesterol in man. *Am J Clin Nutr* 17 (5): 281–295.

Heird, W. C., and A. Lapillonne. 2005. The role of essential fatty acids in development. *Annu Rev Nutr* 25:549–571.

Heller, A., T. Koch, J. Schmeck, and K. van Ackern. 1998. Lipid mediators in inflammatory disorders. *Drugs* 55 (4): 487–496.

Hibbeln, J. R. 2002. Seafood consumption, the DHA content of mothers' milk and prevalence rates of postpartum depression: A cross-national, ecological analysis. *J Affect Disord* 69 (1–3): 15–29.

Hibbeln, J. R., and N. Salem, Jr. 1995. Dietary polyunsaturated fatty acids and depression: When cholesterol does not satisfy. *Am J Clin Nutr* 62 (1): 1–9.

Higgins, H. L. 1916. The rapidity with which alcohol and some sugars may serve as nutrients. *Am J Physiol* 41:258–263.

Ho, L., W. Qin, P. N. Pompl, Z. Xiang, J. Wang, Z. Zhao, Y. Peng, et al. 2004. Diet-induced insulin resistance promotes amyloidosis in a transgenic mouse model of Alzheimer's disease. *FASEB J* 18 (7): 902–904.

Hobhouse, Henry. 1985. Sugar and the slave trade. In *Seeds of change: Five plants that transformed mankind.* London: Sedgwick and Jackson.

Honda, H., Y. Shimizu, and M. Rutter. 2005. No effect of MMR withdrawal on the incidence of autism: A total population study. *J Child Psychol Psychiatry* 46 (6): 572–579.

Horrobin, D. F., and M. S. Manku. 1980. Possible role of prostaglandin E1 in the affective disorders and in alcoholism. *Br Med J* 280 (6228): 1363–1366.

Hosoe, K., M. Kitano, H. Kishida, H. Kubo, K. Fujii, and M. Kitahara. 2007. Study on safety and bioavailability of ubiquinol (Kaneka QH) after single and 4-week multiple oral administration to healthy volunteers. *Regul Toxicol Pharmacol* 47 (1): 19–28.

Huffman, J., and E. H. Kossoff. 2006. State of the ketogenic diet(s) in epilepsy. *Curr Neurol Neurosci Rep* 6 (4): 332–340.

Humphrey, L. L., R. Fu, K. Rogers, M. Freeman, and M. Helfand. 2008. Homocysteine level and coronary heart disease incidence: A systematic review and meta-analysis. *Mayo Clin Proc* 83 (11): 1203–1212.

Huskisson, E. C., H. Berry, P. Gishen, R. W. Jubb, and J. Whitehead. 1995. Effects of antiinflammatory drugs on the progression of osteoarthritis of the knee. LINK Study Group. Longitudinal Investigation of Nonsteroidal Antiinflammatory Drugs in Knee Osteoarthritis. *J Rheumatol* 22 (10): 1941–1946.

Hussain, T., S. Gupta, V. M. Adhami, and H. Mukhtar. 2005. Green tea constituent epigallocatechin-3-gallate selectively inhibits COX-2 without affecting COX-1 expression in human prostate carcinoma cells. *Int J Cancer* 113 (4): 660–669.

Hutter-Paier, B., H. J. Huttunen, L. Puglielli, C. B. Eckman, D. Y. Kim, A. Hofmeister, R. D. Moir, et al. 2004. The ACAT inhibitor CP-113,818 markedly reduces amyloid pathology in a mouse model of Alzheimer's disease. *Neuron* 44 (2): 227–238.

Hyakukoku, M., K. Higashiura, N. Ura, H. Murakami, K. Yamaguchi, L. Wang, M. Furuhashi, N. Togashi, and K. Shimamoto. 2003. Tissue-specific impairment of insulin signaling in vasculature and skeletal muscle of fructose-fed rats. *Hypertens Res* 26 (2): 169–176.

Inoue, M., H. Itoh, T. Tanaka, T. H. Chun, K. Doi, Y. Fukunaga, N. Sawada, et al. 2001. Oxidized LDL regulates vascular endothelial growth factor expression in human macrophages and endothelial cells through activation of peroxisome proliferator-activated receptor-gamma. *Arterioscler Thromb Vasc Biol* 21 (4): 560–566.

Insel, Paul A. 1990. Analgesic-antipyretics and antiinflammatory agents: Drugs employed in the treatment of rheumatoid arthritis and gout. In *Goodman and Gilman's the pharmacological basis of therapeutics*, ed. A. G. Gilman, T. W. Rall, A. S. Nies, and P. Taylor. New York: Pergamon Press.

Ip, C., S. F. Chin, J. A. Scimeca, and M. W. Pariza. 1991. Mammary cancer prevention by conjugated dienoic derivative of linoleic acid. *Cancer Res* 51 (22): 6118–6124.

Jakobisiak, M., and J. Golab. 2003. Potential antitumor effects of statins (review). *Int J Oncol* 23 (4): 1055–1069.

Janniger, C. K., and S. P. Racis. 1987. The arachidonic acid cascade: An immunologically based review. *J Med* 18 (2): 69–80.

Jialal, I., S. Devaraj, and S. K. Venugopal. 2004. C-reactive protein: Risk marker or mediator in atherothrombosis? *Hypertension* 44 (1): 6–11.

Jiang, R., D. C. Paik, J. L. Hankinson, and R. G. Barr. 2007. Cured meat consumption, lung function, and chronic obstructive pulmonary disease among United States adults. *Am J Respir Crit Care Med* 175 (8): 798–804.

Jones, P. J., P. B. Pencharz, and M. T. Clandinin. 1985. Whole body oxidation of dietary fatty acids: Implications for energy utilization. *Am J Clin Nutr* 42 (5): 769–777.

Jump, D. B. 2008. N-3 polyunsaturated fatty acid regulation of hepatic gene transcription. *Curr Opin Lipidol* 19 (3): 242–247.

Juni, P., A. W. Rutjes, and P. A. Dieppe. 2002. Are selective COX 2 inhibitors superior to traditional non steroidal anti-inflammatory drugs? *BMJ* 324 (7349): 1287–1288.

Kaether, C., and C. Haass. 2004. A lipid boundary separates APP and secretases and limits amyloid beta-peptide generation. *J Cell Biol* 167 (5): 809–812.

Kamal-Bahl, S. J., T. Burke, D. Watson, and C. Wentworth. 2007. Discontinuation of lipid modifying drugs among commercially insured United States patients in recent clinical practice. *Am J Cardiol* 99 (4): 530–534.

Kannel, W. B., T. R. Dawber, A. Kagan, N. Revotskie, and J. Stokes. 1961. Factors of risk in the development of coronary heart disease: Six-year follow-up experience. *Ann Intern Med* 55:33–50.

Karg, E., P. Klivenyi, I. Nemeth, K. Bencsik, S. Pinter, and L. Vecsei. 1999. Nonenzymatic antioxidants of blood in multiple sclerosis. *J Neurol* 246 (7): 533–539.

Karmali, R. A. 1987. Eicosanoids in neoplasia. *Prev Med* 16 (4): 493–502.

Kelley, K. A., L. Ho, D. Winger, J. Freire-Moar, C. B. Borelli, P. S. Aisen, and G. M. Pasinetti. 1999. Potentiation of excitotoxicity in transgenic mice overexpressing neuronal cyclo-oxygenase-2. *Am J Pathol* 155 (3): 995–1004.

Keys, A. 1957. Diet and the epidemiology of coronary heart disease. *JAMA* 164 (17): 1912–1919.

———. 1980. *Seven countries: A multivariate analysis of death and coronary heart disease.* Cambridge, MA: Harvard University Press.

Keys, Ancel and Margaret. 1975. *How to eat well and stay well the Mediterranean way.* Garden City, N.Y.: Doubleday.

Kim, J., and B. E. Alger. 2004. Inhibition of cyclooxygenase-2 potentiates retrograde endo-cannabinoid effects in hippocampus. *Nat Neurosci* 7 (7): 697–698.

Kirschenbauer, H. G. 1960. In *Fats and oils: An outline of their chemistry and technology.* New York: Reinhold Publishing Corp.

Kitajka, K., L. G. Puskas, A. Zvara, L. Hackler, Jr., G. Barcelo-Coblijn, Y. K. Yeo, and T. Farkas. 2002. The role of n-3 polyunsaturated fatty acids in brain: Modulation of rat brain gene expression by dietary n-3 fatty acids. *Proc Natl Acad Sci USA* 99 (5): 2619–2624.

Kitamura, Y., S. Shimohama, H. Koike, J. Kakimura, Y. Matsuoka, Y. Nomura, P. J. Gebicke-Haerter, and T. Taniguchi. 1999. Increased expression of cyclooxygenases and peroxisome proliferator-activated receptor-gamma in Alzheimer's disease brains. *Biochem Biophys Res Commun* 254 (3): 582–586.

Klenke, F. M., M. M. Gebhard, V. Ewerbeck, A. Abdollahi, P. E. Huber, and A. Sckell. 2006. The selective Cox-2 inhibitor Celecoxib suppresses angiogenesis and growth of secondary bone tumors: An intravital microscopy study in mice. *BMC Cancer* 6:9.

Klerman, G. L., and M. M. Weissman. 1989. Increasing rates of depression. *JAMA* 261 (15): 2229–2235.

Knijff, P. De, A. M. van den Maagdenberg, R. R. Frants, and L. M. Havekes. 1994. Genetic heterogeneity of apolipoprotein E and its influence on plasma lipid and lipoprotein levels. *Hum Mutat* 4:178–194.

Knott, C., G. Stern, and G. P. Wilkin. 2000. Inflammatory regulators in Parkinson's disease: iNOS, lipocortin-1, and cyclooxygenases-1 and -2. *Mol Cell Neurosci* 16 (6): 724–739.

Koenig, W., and M. B. Pepys. 2002. C-reactive protein risk prediction: Low specificity, high sensitivity. *Ann Intern Med* 136 (7): 550–552.

Kohlmeier, L., N. Simonsen, P. van 't Veer, J. J. Strain, J. M. Martin-Moreno, B. Margolin, J. K. Huttunen, et al. 1997. Adipose tissue trans fatty acids and breast cancer in the European Community Multicenter Study on Antioxidants, Myocardial Infarction, and Breast Cancer. *Cancer Epidemiol Biomarkers Prev* 6 (9): 705–710.

Kojro, E., G. Gimpl, S. Lammich, W. Marz, and F. Fahrenholz. 2001. Low cholesterol stimulates the nonamyloidogenic pathway by its effect on the alpha-secretase ADAM 10. *Proc Natl Acad Sci USA* 98 (10): 5815–5820.

Kritchevsky, D. 1998. History of recommendations to the public about dietary fat. *J Nutr* 128 (2 Suppl): 449S–452S.

Kronmal, R. A., R. G. Hart, T. A. Manolio, R. L. Talbert, N. J. Beauchamp, and A. Newman. 1998. Aspirin use and incident stroke in the cardiovascular health study. CHS Collab-orative Research Group. *Stroke* 29 (5): 887–894.

Kuczynski, B., and N. V. Reo. 2006. Evidence that plasmalogen is protective against oxidative stress in the rat brain. *Neurochem Res* 31 (5): 639–656.

Kummerow, F. A., Y. Kim, J. Hull, J. Pollard, P. Ilinov, D. L. Drossiev, and J. Valek. 1977. The influence of egg consumption on the serum cholesterol level in human subjects. *Am J Clin Nutr* 30 (5): 664–673.

Kump, L. R., J. F. Kasting, and R. G. Crane. 1999. *The earth system*. Upper Saddle River, NJ: Prentice-Hall.

Kurzok, R., and C. C. Lieb. 1930. Biochemical studies of human semen: The action of semen on the human uterus. *Proc Soc Exp Biol Med* 28:268–272.

Ladwig, K. H., B. Marten-Mittag, H. Lowel, A. Doring, and W. Koenig. 2005. C-reactive protein, depressed mood, and the prediction of coronary heart disease in initially healthy men: Results from the MONICA-KORA Augsburg Cohort Study 1984–1998. *Eur Heart J* 26 (23): 2537–2542.

Ladwig, K. H., B. Marten-Mittag, H. Lowel, A. Doring, and H. E. Wichmann. 2006. Synergistic effects of depressed mood and obesity on long-term cardiovascular risks in 1510 obese men and women: Results from the MONICA-KORA Augsburg Cohort Study 1984–1998. *Int J Obes* (Lond) 30 (9): 1408–1414.

Landers, Peter. 2006. Stalking cholesterol: How one scientist intrigued by molds found first statin. *Wall Street Journal*, January 9, 2006, A1.

Lands, W. E., T. Hamazaki, K. Yamazaki, H. Okuyama, K. Sakai, Y. Goto, and V. S. Hubbard. 1990. Changing dietary patterns. *Am J Clin Nutr* 51 (6): 991–993.

Langston, J. W., P. Ballard, J. W. Tetrud, and I. Irwin. 1983. Chronic parkinsonism in humans due to a product of meperidine-analog synthesis. *Science* 219 (4587): 979–980.

Larsson, S. C., M. Kumlin, M. Ingelman-Sundberg, and A. Wolk. 2004. Dietary long-chain n-3 fatty acids for the prevention of cancer: A review of potential mechanisms. *Am J Clin Nutr* 79 (6): 935–945.

Lauritzen, L., H. S. Hansen, M. H. Jorgensen, and K. F. Michaelsen. 2001. The essentiality of long chain n-3 fatty acids in relation to development and function of the brain and retina. *Prog Lipid Res* 40 (1–2): 1–94.

Law, M. R., N. J. Wald, and A. R. Rudnicka. 2003. Quantifying effect of statins on low density lipoprotein cholesterol, ischaemic heart disease, and stroke: Systematic review and meta-analysis. *BMJ* 326 (7404): 1423–1427.

Lawrence, G. D. 1990. Effect of dietary lipids on adjuvant-induced arthritis in rats. *Nutr Res* 10:283–290.

Lawrence, G. D., G. Cohen, A. H. Aufses, and G. I. Slater. 1984. Lipid peroxidation in dimethylhydrazine-induced colon cancer. *Ann NY Acad Sci* 435:441–443.

Lawrence, G. D., A. Mavi, and K. Meral. 2008. Promotion by phosphate of Fe(III)– and Cu(II)–catalyzed autoxidation of fructose. *Carbohydr Res* 343 (4): 626–635.

Lebeau, A., F. Terro, W. Rostene, and D. Pelaprat. 2004. Blockade of 12-lipoxygenase expression protects cortical neurons from apoptosis induced by beta-amyloid peptide. *Cell Death Differ* 11 (8): 875–884.

Lee, Y. S., C. H. Bailey, E. R. Kandel, and B. K. Kaang. 2008. Transcriptional regulation of long-term memory in the marine snail *Aplysia*. *Mol Brain* 1 (1): 3.

Libby, P. 2000. Changing concepts of atherogenesis. *J Intern Med* 247 (3): 349–358.

LIPID. 1998. Prevention of cardiovascular events and death with pravastatin in patients with coronary heart disease and a broad range of initial cholesterol levels. The Long-Term Intervention with Pravastatin in Ischaemic Disease (LIPID) Study Group. *N Engl J Med* 339 (19): 1349–1357.

Lipid Research Clinics Study Group. 1984. The Lipid Research Clinics Coronary Primary Prevention Trial results. I. Reduction in incidence of coronary heart disease. *JAMA* 251 (3): 351–364.

Liu, N.S.T., and O. A. Roels. 1980. The vitamins. In *Modern nutrition in health and disease*, ed. R.S.Goodhart and M. E. Shils. Philadelphia: Lea and Febiger.

Lock, A. L., B. A. Corl, D. M. Barbano, D. E. Bauman, and C. Ip. 2004. The anticarcinogenic effect of trans-11 18:1 is dependent on its conversion to cis-9, trans-11 CLA by delta9-desaturase in rats. *J Nutr* 134 (10): 2698–2704.

Lukiw, W. J., J. G. Cui, V. L. Marcheselli, M. Bodker, A. Botkjaer, K. Gotlinger, C. N. Serhan, and N. G. Bazan. 2005. A role for docosahexaenoic acid-derived neuroprotectin D1 in neural cell survival and Alzheimer disease. *J Clin Invest* 115 (10): 2774–2783.

Lynn, A. B., and M. Herkenham. 1994. Localization of cannabinoid receptors and nonsaturable high-density cannabinoid binding sites in peripheral tissues of the rat: Implications for receptor-mediated immune modulation by cannabinoids. *J Pharmacol Exp Ther* 268 (3): 1612–1623.

Maccarone, M., N. Battista, and D. Centonze. 2007. The endocannabinoid pathway in Huntington's disease: A comparison with other neurodegenerative diseases. *Prog Neurobiol* 81 (5–6): 349–379.

Macdonald, I. 1966. Influence of fructose and glucose on serum lipid levels in men and pre- and postmenopausal women. *Am J Clin Nutr* 18 (5): 369–372.

Macdonald, I., A. Keyser, and D. Pacy. 1978. Some effects, in man, of varying the load of glucose, sucrose, fructose, or sorbitol on various metabolites in blood. *Am J Clin Nutr* 31 (8): 1305–1311.

Mackenzie, I. R., and D. G. Munoz. 1998. Nonsteroidal anti-inflammatory drug use and Alzheimer-type pathology in aging. *Neurology* 50 (4): 986–990.

Maffei, M., J. Halaas, E. Ravussin, R. E. Pratley, G. H. Lee, Y. Zhang, H. Fei, et al. 1995. Leptin levels in human and rodent: Measurement of plasma leptin and ob RNA in obese and weight-reduced subjects. *Nat Med* 1 (11): 1155–1161.

Makrides, M., R. A. Gibson, T. Udell, and K. Ried. 2005. Supplementation of infant formula with long-chain polyunsaturated fatty acids does not influence the growth of term infants. *Am J Clin Nutr* 81 (5): 1094–1101.

Malmros, H., and G. Wigand. 1957. The effect on serum-cholesterol of diets containing different fats. *Lancet* 273 (6984): 1–8.

Mandel, S., E. Grunblatt, P. Riederer, M. Gerlach, Y. Levites, and M. B. Youdim. 2003. Neuroprotective strategies in Parkinson's disease: An update on progress. *CNS Drugs* 17 (10): 729–762.

McCann, J. C., and B. N. Ames. 2005. Is docosahexaenoic acid, an n-3 long-chain polyunsaturated fatty acid, required for development of normal brain function? An overview of evidence from cognitive and behavioral tests in humans and animals. *Am J Clin Nutr* 82 (2): 281–295.

McCully, K. S. 1998. Homocysteine, folate, vitamin B6, and cardiovascular disease. *JAMA* 279 (5): 392–393.

McDermott, C. M., G. J. LaHoste, C. Chen, A. Musto, N. G. Bazan, and J. C. Magee. 2003. Sleep deprivation causes behavioral, synaptic, and membrane excitability alterations in hippocampal neurons. *J Neurosci* 23 (29): 9687–9695.

McGarry, J. Denis. 2002. Lipid metabolism I: Utilization and storage of energy in lipid form. In *Textbook of biochemistry with clinical correlations*, ed. T. M. Devlin. New York: Wiley-Liss.

McGeer, P. L., M. Schulzer, and E. G. McGeer. 1996. Arthritis and anti-inflammatory agents as possible protective factors for Alzheimer's disease: A review of 17 epidemiologic studies. *Neurology* 47 (2): 425–432.

McGiff, J. C. 1987. Arachidonic acid metabolites. *Prev Med* 16:503–509.

McKinnon, H., E. Gherardi, M. Reidy, and D. Bowyer. 2006. Hepatocyte growth factor/scatter factor and MET are involved in arterial repair and atherogenesis. *Am J Pathol* 168 (1): 340–348.

Mechoulam, R., L. Hanus, and B. R. Martin. 1994. Search for endogenous ligands of the cannabinoid receptor. *Biochem Pharmacol* 48 (8): 1537–1544.

Medina, E. A., W. F. Horn, N. L. Keim, P. J. Havel, P. Benito, D. S. Kelley, G. J. Nelson, and K. L. Erickson. 2000. Conjugated linoleic acid supplementation in humans: Effects on circulating leptin concentrations and appetite. *Lipids* 35 (7): 783–788.

Melamed, J., J. M. Einhorn, and M. M. Ittmann. 1997. Allelic loss on chromosome 13q in human prostate carcinoma. *Clin Cancer Res* 3 (10): 1867–1872.

Menotti, A., H. Blackburn, D. Kromhout, A. Nissinen, H. Adachi, and M. Lanti. 2001. Cardiovascular risk factors as determinants of 25-year all-cause mortality in the seven countries study. *Eur J Epidemiol* 17 (4): 337–346.

Menzies, G. 2003. *1421: The year China discovered America*. New York: Perennial/Harper-Collins Publishers.

Miller, G. 2003. Society for Neuroscience meeting: Neurons get connected via glia. *Science* 302 (5649): 1323.

Miller, G. W. 2007. Paraquat: The red herring of Parkinson's disease research. *Toxicol Sci* 100 (1): 1–2.

Minghetti, L. 2004. Cyclooxygenase-2 (COX-2) in inflammatory and degenerative brain diseases. *J Neuropathol Exp Neurol* 63 (9): 901–910.

Mishra, S. K., S. C. Watkins, and L. M. Traub. 2002. The autosomal recessive hypercholesterolemia (ARH) protein interfaces directly with the clathrin-coat machinery. *Proc Natl Acad Sci USA* 99 (25): 16099–16104.

Mitchell, E. A., M. G. Aman, S. H. Turbott, and M. Manku. 1987. Clinical characteristics and serum essential fatty acid levels in hyperactive children. *Clin Pediatr* (Phila) 26 (8): 406–411.

Mitchell, J. A., P. Akarasereenont, C. Thiemermann, R. J. Flower, and J. R. Vane. 1993. Selectivity of nonsteroidal antiinflammatory drugs as inhibitors of constitutive and inducible cyclooxygenase. *Proc Natl Acad Sci USA* 90 (24): 11693–11697.

Mizuno, Y., and K. Ohta. 1986. Regional distributions of thiobarbituric acid-reactive products, activities of enzymes regulating the metabolism of oxygen free radicals, and some of the related enzymes in adult and aged rat brains. *J Neurochem* 46 (5): 1344–1352.

Moisse, K., and M. J. Strong. 2006. Innate immunity in amyotrophic lateral sclerosis. *Biochim Biophys Acta* 1762 (11–12): 1083–1093.

Morel, D. W., J. R. Hessler, and G. M. Chisolm. 1983. Low density lipoprotein cytotoxicity induced by free radical peroxidation of lipid. *J Lipid Res* 24 (8): 1070–1076.

Mori, T. A., and R. J. Woodman. 2006. The independent effects of eicosapentaenoic acid and docosahexaenoic acid on cardiovascular risk factors in humans. *Curr Opin Clin Nutr Metab Care* 9 (2): 95–104.

MSNBC. 2006. New York City passes trans fat ban [accessed June 30, 2009]. http://www.msnbc.msn.com/id/16051436/.

Mudd, S. H., F. Skovby, H. L. Levy, K. D. Pettigrew, B. Wilcken, R. E. Pyeritz, G. Andria, et al. 1985. The natural history of homocysteine due to cystathionine beta synthase deficiency. *Am J Hum Gen* 37:1–31.

Multiple Risk Factor Intervention Trial Research Group. 1982. Multiple Risk Factor Intervention Trial: Risk factor changes and mortality results. Multiple Risk Factor Intervention Trial Research Group. *JAMA* 248 (12): 1465–1477.

Murch, S. J., P. A. Cox, S. A. Banack, J. C. Steele, and O. W. Sacks. 2004. Occurrence of betamethylamino-1 -alanine (BMAA) in ALS/PDC patients from Guam. *Acta Neurol Scand* 110 (4): 267–269.

Nagakura, T., S. Matsuda, K. Shichijyo, H. Sugimoto, and K. Hata. 2000. Dietary supplementation with fish oil rich in omega-3 polyunsaturated fatty acids in children with bronchial asthma. *Eur Respir J* 16 (5): 861–865.

Nagan, N., and R. A. Zoeller. 2001. Plasmalogens: Biosynthesis and functions. *Prog Lipid Res* 40 (3): 199–229.

Nauss, K. M., M. Locniskar, and P. M. Newberne. 1983. Effect of alterations in the quality and quantity of dietary fat on 1,2-dimethylhydrazine-induced colon tumorigenesis in rats. *Cancer Res* 43 (9): 4083–4090.

Navab, M., G. M. Ananthramaiah, S. T. Reddy, B. J. Van Lenten, B. J. Ansell, G. C. Fonarow, K. Vahabzadeh, et al. 2004. The oxidation hypothesis of atherogenesis: The role of oxidized phospholipids and HDL. *J Lipid Res* 45 (6): 993–1007.

Ng, T. K., K. C. Hayes, G. F. DeWitt, M. Jegathesan, N. Satgunasingam, A. S. Ong, and D. Tan. 1992. Dietary palmitic and oleic acids exert similar effects on serum cholesterol and lipoprotein profiles in normocholesterolemic men and women. *J Am Coll Nutr* 11 (4): 383–390.

Nishigaki, I., M. Hagihara, H. Tsunekawa, M. Maseki, and K. Yagi. 1981. Lipid peroxide levels of serum lipoprotein fractions of diabetic patients. *Biochem Med* 25 (3): 373–378.

Norat, T., S. Bingham, P. Ferrari, N. Slimani, M. Jenab, M. Mazuir, K. Overvad, et al. 2005. Meat, fish, and colorectal cancer risk: The European Prospective Investigation into cancer and nutrition. *J Natl Cancer Inst* 97 (12): 906–916.

Okamoto, K., T. Kihira, T. Kondo, G. Kobashi, M. Washio, S. Sasaki, T. Yokoyama, et al. 2007. Nutritional status and risk of amyotrophic lateral sclerosis in Japan. *Amyotroph Lateral Scler* 8 (5): 300–304.

Okuyama, H., T. Kobayashi, and S. Watanabe. 1996. Dietary fatty acids—the N-6/N-3 balance and chronic elderly diseases: Excess linoleic acid and relative N-3 deficiency syndrome seen in Japan. *Prog Lipid Res* 35 (4): 409–457.

Oliveros, J. C., M. K. Jandali, M. Timsit-Berthier, R. Remy, A. Benghezal, A. Audibert, and J. M. Moeglen. 1978. Vasopressin in amnesia. *Lancet* 311 (8054): 42.

Olsen, S. F., P. Grandjean, P. Weihe, and T. Videro. 1993. Frequency of seafood intake in pregnancy as a determinant of birth weight: Evidence for a dose dependent relationship. *J Epidemiol Community Health* 47 (6): 436–440.

Olsen, S. F., and N. J. Secher. 2002. Low consumption of seafood in early pregnancy as a risk factor for preterm delivery: Prospective cohort study. *BMJ* 324 (7335): 447.

Oteiza, P. I., O. D. Uchitel, F. Carrasquedo, A. L. Dubrovski, J. C. Roma, and C. G. Fraga. 1997. Evaluation of antioxidants, protein, and lipid oxidation products in blood from sporadic amyotrophic lateral sclerosis patients. *Neurochem Res* 22 (4): 535–539.

Palinski, W., M. E. Rosenfeld, S. Yla-Herttuala, G. C. Gurtner, S. S. Socher, S. W. Butler, S. Parthasarathy, T. E. Carew, D. Steinberg, and J. L. Witztum. 1989. Low density lipoprotein undergoes oxidative modification in vivo. *Proc Natl Acad Sci USA* 86 (4): 1372–1376.

Panagiotakos, D. B., C. Pitsavos, C. Chrysohoou, S. Kavouras, and C. Stefanadis. 2005. The associations between leisure-time physical activity and inflammatory and coagulation markers related to cardiovascular disease: The ATTICA study. *Prev Med* 40 (4): 432–437.

Panagiotakos, D. B., C. Pitsavos, C. Chrysohoou, E. Tsetsekou, C. Papageorgiou, G. Christodoulou, and C. Stefanadis. 2004. Inflammation, coagulation, and depressive symptomatology in cardiovascular disease-free people: The ATTICA study. *Eur Heart J* 25 (6): 492–499.

Paoletti, R., C. Bolego, and A. Cignarella. 2005. Lipid and non-lipid effects of statins. *Handb Exp Pharmacol* (170): 365–388.

Parker, G., N. A. Gibson, H. Brotchie, G. Heruc, A. M. Rees, and D. Hadzi-Pavlovic. 2006. Omega-3 fatty acids and mood disorders. *Am J Psychiatry* 163 (6): 969–978.

Pasinetti, G. M. 1998. Cyclooxygenase and inflammation in Alzheimer's disease: Experimental approaches and clinical interventions. *J Neurosci Res* 54 (1): 1–6.

Pauling, Linus. 1970. *Vitamin C and the common cold.* San Fransisco: W. H. Freeman and Co.

Payne, A. 2001. Nutrition and diet in the clinical management of multiple sclerosis. *J Hum Nutr Diet* 14 (5): 349–357.

Pearce, M. L., and S. Dayton. 1971. Incidence of cancer in men on a diet high in poly-unsaturated fat. *Lancet* 297 (7697): 464–467.

Peet, M. 2004. Nutrition and schizophrenia: Beyond omega-3 fatty acids. *Prostaglandins Leukot Essent Fatty Acids* 70 (4): 417–422.

Pekkanen, J., A. Nissinen, S. Punsar, and M. J. Karvonen. 1989. Serum cholesterol and risk of accidental or violent death in a 25-year follow-up: The Finnish cohorts of the Seven Countries Study. *Arch Intern Med* 149 (7): 1589–1591.

Peterson, A. M., and W. F. McGhan. 2005. Pharmacoeconomic impact of non-compliance with statins. *Pharmacoeconomics* 23 (1): 13–25.

Pirozzo, S., C. Summerbell, C. Cameron, and P. Glasziou. 2002. Advice on low-fat diets for obesity. *Cochrane Database Syst Rev* (2):CD003640.

Pollan, Michael. 2001. *The botany of desire: A plant's eye-view of the world.* New York: Random House.

Porter, A. C., J. M. Sauer, M. D. Knierman, G. W. Becker, M. J. Berna, J. Bao, G. G. Nomikos, et al. 2002. Characterization of a novel endocannabinoid, virodhamine, with antagonist activity at the CB1 receptor. *J Pharmacol Exp Ther* 301 (3): 1020–1024.

Pozuelo, L., G. Tesar, J. Zhang, M. Penn, K. Franco, and W. Jiang. 2009. Depression and heart disease: What do we know, and where are we headed? *Cleve Clin J Med* 76 (1): 59–70.

Pryor, W. A. 2000. Vitamin E and heart disease: Basic science to clinical intervention trials. *Free Radic Biol Med* 28 (1): 141–164.

Puri, B. K., S. J. Counsell, G. Hamilton, A. J. Richardson, and D. F. Horrobin. 2001. Eicosap-entaenoic acid in treatment-resistant depression associated with symptom remission, structural brain changes and reduced neuronal phospholipid turnover. *Int J Clin Pract* 55 (8): 560–563.

Putnam, J. J., and J. E. Allshouse. 1999. Food consumption, prices and expenditures, 1970–1997. Statistical Bulletin No. 965. U.S. Department of Agriculture, Economic Research Services [accessed June 30, 2009]. http://usda.mannlib.cornell.edu/usda/reports/general/sb/sb965.pdf.

Quinn, M. T., S. Parthasarathy, L. G. Fong, and D. Steinberg. 1987. Oxidatively modified low density lipoproteins: A potential role in recruitment and retention of monocyte/macro-phages during atherogenesis. *Proc Natl Acad Sci USA* 84 (9): 2995–2998.

Randall, M. D., D. A. Kendall, and S. O'Sullivan. 2004. The complexities of the cardiovascular actions of cannabinoids. *Br J Pharmacol* 142 (1): 20–26.

Ravnskov, U., P. J. Rosch, and M. C. Sutter. 2005. Intensive lipid lowering with atorvastatin in coronary disease. *N Engl J Med* 353 (1): 93–96; author reply 93–96.

Reaven, P., S. Parthasarathy, B. J. Grasse, E. Miller, D. Steinberg, and J. L. Witztum. 1993. Effects of oleate-rich and linoleate-rich diets on the susceptibility of low density lipoprotein to oxidative modification in mildly hypercholesterolemic subjects. *J Clin Invest* 91 (2): 668–676.

Reszko, A. E., T. Kasumov, F. David, K. A. Jobbins, K. R. Thomas, C. L. Hoppel, H. Brunengraber, and C. Des Rosiers. 2004. Peroxisomal fatty acid oxidation is a substantial source of the acetyl moiety of malonyl-CoA in rat heart. *J Biol Chem* 279 (19): 19574–19579.

Richard, D., K. Kefi, U. Barbe, P. Bausero, and F. Visioli. 2008. Polyunsaturated fatty acids as antioxidants. *Pharmacol Res* 57 (6): 451–455.

Richardson, A. J., C. M. Calvin, C. Clisby, D. R. Schoenheimer, P. Montgomery, J. A. Hall, G. Hebb, E. Westwood, J. B. Talcott, and J. F. Stein. 2000. Fatty acid deficiency signs predict the severity of reading and related difficulties in dyslexic children. *Prostaglandins Leukot Essent Fatty Acids* 63 (1–2): 69–74.

Richardson, A. J., I. J. Cox, J. Sargentoni, and B. K. Puri. 1997. Abnormal cerebral phospholipid metabolism in dyslexia indicated by phosphorus-31 magnetic resonance spectroscopy. *NMR Biomed* 10 (7): 309–314.

Ridker, P. M., J. E. Buring, N. R. Cook, and N. Rifai. 2003. C-reactive protein, the metabolic syndrome, and risk of incident cardiovascular events: An 8-year follow-up of 14 719 initially healthy American women. *Circulation* 107 (3) :391–397.

Ridker, P. M., E. Danielson, F. A. Fonseca, J. Genest, A. M. Gotto, Jr., J. J. Kastelein, W. Koenig, et al. 2008. Rosuvastatin to prevent vascular events in men and women with elevated C-reactive protein. *N Engl J Med* 359 (21): 2195–2207.

Rolls, B. J., E. L. Morris, and L. S. Roe. 2002. Portion size of food affects energy intake in normal-weight and overweight men and women. *Am J Clin Nutr* 76 (6): 1207–1213.

Rondeau, V. 2002. A review of epidemiologic studies on aluminum and silica in relation to Alzheimer's disease and associated disorders. *Rev Environ Health* 17 (2): 107–121.

Rong, J. X., S. Rangaswamy, L. Shen, R. Dave, Y. H. Chang, H. Peterson, H. N. Hodis, G. M. Chisolm, and A. Sevanian. 1998. Arterial injury by cholesterol oxidation products causes endothelial dysfunction and arterial wall cholesterol accumulation. *Arterioscler Thromb Vasc Biol* 18 (12): 1885–1894.

Ross, B. M., I. McKenzie, I. Glen, and C. P. Bennett. 2003. Increased levels of ethane, a non-invasive marker of n-3 fatty acid oxidation, in breath of children with attention deficit hyperactivity disorder. *Nutr Neurosci* 6 (5): 277–281.

Roth, G. J., and P. W. Majerus. 1975. The mechanism of the effect of aspirin on human platelets. I. Acetylation of a particulate fraction protein. *J Clin Invest* 56 (3): 624–632.

Roux, L., M. Pratt, T. O. Tengs, M. M. Yore, T. L. Yanagawa, J. Van Den Bos, C. Rutt, R. C. Brownson, K. E. Powell, G. Heath, H. W. Kohl, 3rd, S. Teutsch, J. Cawley, I. M. Lee, L. West, and D. M. Buchner. 2008. Cost effectiveness of community-based physical activity interventions. *Am J Prev Med* 35 (6): 578–588.

Rudin, D. O. 1981. The major psychoses and neuroses as omega-3 essential fatty acid deficiency syndrome: Substrate pellagra. *Biol Psychiatry* 16 (9): 837–850.

Ruyter, B., C. Rosjo, B. Grisdale-Helland, G. Rosenlund, A. Obach, and M. S. Thomassen. 2003. Influence of temperature and high dietary linoleic acid content on esterification, elongation, and desaturation of PUFA in Atlantic salmon hepatocytes. *Lipids* 38 (8): 833–840.

Sagredo, O., J. A. Ramos, A. Decio, R. Mechoulam, and J. Fernandez-Ruiz. 2007. Cannabidiol reduced the striatal atrophy caused 3-nitropropionic acid in vivo by mechanisms independent of the activation of cannabinoid, vanilloid TRPV1 and adenosine A2A receptors. *Eur J Neurosci* 26 (4): 843–851.

Sakai, M., M. Oimomi, and M. Kasuga. 2002. Experimental studies on the role of fructose in the development of diabetic complications. *Kobe J Med Sci* 48 (5–6): 125–136.

Samaha, F. F., G. D. Foster, and A. P. Makris. 2007. Low-carbohydrate diets, obesity, and metabolic risk factors for cardiovascular disease. *Curr Atheroscler Rep* 9 (6): 441–447.

Samaha, F. F., N. Iqbal, P. Seshadri, K. L. Chicano, D. A. Daily, J. McGrory, T. Williams, M. Williams, E. J. Gracely, and L. Stern. 2003. A low-carbohydrate as compared with a low-fat diet in severe obesity. *N Engl J Med* 348 (21): 2074–2081.

Sanak, M., B. D. Levy, C. B. Clish, N. Chiang, K. Gronert, L. Mastalerz, C. N. Serhan, and A. Szczeklik. 2000. Aspirin-tolerant asthmatics generate more lipoxins than aspirin-intolerant asthmatics. *Eur Respir J* 16 (1): 44–49.

Sanda, Bill. 2004. The double danger of high fructose corn syrup. Weston A. Price Foundation, February 19, 2004 [accessed June 30, 2009]. http://www.westonaprice. org/modernfood/highfrucose.html.

Sanders, T. A. 1999. Essential fatty acid requirements of vegetarians in pregnancy, lactation, and infancy. *Am J Clin Nutr* 70 (3 Suppl): 555S–559S.

Sato, Y., N. Hotta, N. Sakamoto, S. Matsuoka, N. Ohishi, and K. Yagi. 1979. Lipid peroxide level in plasma of diabetic patients. *Biochem Med* 21 (1): 104–107.

Scandinavian Simvastatin Survival Study. 1994. Randomised trial of cholesterol lowering in 4444 patients with coronary heart disease. *Lancet* 344 (8934): 1383–1389.

Schlanger, S., M. Shinitzky, and D. Yam. 2002. Diet enriched with omega-3 fatty acids alleviates convulsion symptoms in epilepsy patients. *Epilepsia* 43 (1): 103–104.

Schmitz, G., and J. Ecker. 2008. The opposing effects of n-3 and n-6 fatty acids. *Prog Lipid Res* 47 (2): 147–155.

Seierstad, S. L., I. Seljeflot, O. Johansen, R. Hansen, M. Haugen, G. Rosenlund, L. Froyland, and H. Arnesen. 2005. Dietary intake of differently fed salmon: The influence on markers of human atherosclerosis. *Eur J Clin Invest* 35 (1): 52–59.

Serhan, C. N. 2001. Lipoxins and aspirin-triggered 15-epi-lipoxins are endogenous components of antiinflammation: Emergence of the counterregulatory side. *Arch Immunol Ther Exp* (Warsz) 49 (3): 177–188.

Serhan, C. N., M. Arita, S. Hong, and K. Gotlinger. 2004. Resolvins, docosatrienes, and neuroprotectins, novel omega-3-derived mediators, and their endogenous aspirin-triggered epimers. *Lipids* 39 (11): 1125–1132.

Serhan, C. N., S. Hong, K. Gronert, S. P. Colgan, P. R. Devchand, G. Mirick, and R. L. Moussignac. 2002. Resolvins: A family of bioactive products of omega-3 fatty acid transformation circuits initiated by aspirin treatment that counter proinflammation signals. *J Exp Med* 196 (8): 1025–1037.

Serhan, C. N., and E. Oliw. 2001. Unorthodox routes to prostanoid formation: New twists in cyclooxygenase-initiated pathways. *J Clin Invest* 107 (12): 1481–1489.

Shafiq, N., S. Malhotra, P. Pandhi, and A. Grover. 2005. The "statinth" wonder of the world: A panacea for all illnesses or a bubble about to burst. *J Negat Results Biomed* 4:3.

Shaftel, S. S., J. A. Olschowka, S. D. Hurley, A. H. Moore, and M. K. O'Banion. 2003. COX-3: A splice variant of cyclooxygenase-1 in mouse neural tissue and cells. *Brain Res Mol Brain Res* 119 (2): 213–215.

Sharman, I. M. 1981. Glycogen loading: Advantages but possible disadvantages. *Br J Sports Med* 15:64–67.

Sheaff Greiner, R. C., Q. Zhang, K. J. Goodman, D. A. Giussani, P. W. Nathanielsz, and J. T. Brenna. 1996. Linoleate, alpha-linolenate, and docosahexaenoate recycling into saturated and monounsaturated fatty acids is a major pathway in pregnant or lactating adults and fetal or infant rhesus monkeys. *J Lipid Res* 37 (12): 2675–2686.

Sheng, H., J. Shao, J. D. Morrow, R. D. Beauchamp, and R. N. DuBois. 1998. Modulation of apoptosis and Bcl-2 expression by prostaglandin E2 in human colon cancer cells. *Cancer Res* 58 (2): 362–366.

Sheng, M., and M. J. Kim. 2002. Postsynaptic signaling and plasticity mechanisms. *Science* 298 (5594): 776–780.

Shepherd, J., S. M. Cobbe, I. Ford, C. G. Isles, A. R. Lorimer, P. W. MacFarlane, J. H. McKillop, and C. J. Packard. 1995. Prevention of coronary heart disease with pravastatin in men with hypercholesterolemia. West of Scotland Coronary Prevention Study Group. *N Engl J Med* 333 (20): 1301–1307.

Sherwood, Lauralee. 2004. *Human physiology: From cells to systems.* 5th ed. Belmont, CA: Brooks/Cole-Thomson Learning.

Shimizu, I., Y. R. Ma, Y. Mizobuchi, F. Liu, T. Miura, Y. Nakai, M. Yasuda, et al. 1999. Effects of Sho-saiko-to, a Japanese herbal medicine, on hepatic fibrosis in rats. *Hepatology* 29 (1): 149–160.

Silva, M., M. L. Matthews, C. Jarvis, N. M. Nolan, P. Belliveau, M. Malloy, and P. Gandhi. 2007. Meta-analysis of drug-induced adverse events associated with intensive-dose statin therapy. *Clin Ther* 29 (2): 253–260.

Simon, A. M., M. B. Manigrasso, and J. P. O'Connor. 2002. Cyclo-oxygenase 2 function is essential for bone fracture healing. *J Bone Miner Res* 17 (6): 963–976.

Simons, K., and D. Toomre. 2000. Lipid rafts and signal transduction. *Nat Rev Mol Cell Biol* 1 (1): 31–39.

Simopoulos, A. P. 1991. Omega-3 fatty acids in health and disease and in growth and development. *Am J Clin Nutr* 54 (3): 438–463.

———. 2002a. The importance of the ratio of omega-6/omega-3 essential fatty acids. *Biomed Pharmacother* 56 (8): 365–379.

———. 2002b. Omega-3 fatty acids in inflammation and autoimmune diseases. *J Am Coll Nutr* 21 (6): 495–505.

Sinclair, A. J., D. Begg, M. Mathai, and R. S. Weisinger. 2007. Omega 3 fatty acids and the brain: Review of studies in depression. *Asia Pac J Clin Nutr* 16 (Suppl 1): 391–397.

Sing, C. F., and E. A. Boerwinkle. 1987. Genetic architecture of inter-individual variability in apolipoprotein, lipoprotein and lipid phenotypes. *Ciba Found Symp* 130:99–127.

Singh, M. 2005. Essential fatty acids, DHA and human brain. *Indian J Pediatr* 72 (3): 239–242.

Slaga, T. J., A. J. Klein-Szanto, L. L. Triplett, L. P. Yotti, and K. E. Trosko. 1981. Skin tumor-promoting activity of benzoyl peroxide, a widely used free radical-generating compound. *Science* 213 (4511): 1023–1025.

Smith, S. C., Jr. 1996. Risk-reduction therapy: The challenge to change. Presented at the 68th scientific sessions of the American Heart Association, November 13, 1995, Anaheim, CA. *Circulation* 93 (12): 2205–2211.

Smuts, C. M., M. Huang, D. Mundy, T. Plasse, S. Major, and S. E. Carlson. 2003. A randomized trial of docosahexaenoic acid supplementation during the third trimester of pregnancy. *Obstet Gynecol* 101 (3): 469–479.

Snyderman, S. E. 1980. Nutrition in infancy and adolescence. In *Modern nutrition in health and disease*, ed. R. S. Goodhart and M. E. Shils. Philadelphia: Lea and Febiger.

Sohal, R. S., and M. J. Forster. 2007. Coenzyme Q, oxidative stress and aging. *Mitochondrion* 7 (Suppl): S103–S111.

Sohal, R. S., S. Kamzalov, N. Sumien, M. Ferguson, I. Rebrin, K. R. Heinrich, and M. J. Forster. 2006. Effect of coenzyme Q10 intake on endogenous coenzyme Q content, mitochondrial electron transport chain, antioxidative defenses, and life span of mice. *Free Radic Biol Med* 40 (3): 480–487.

Song, Y. G., H. M. Kwon, J. M. Kim, B. K. Hong, D. S. Kim, A. J. Huh, K. H. Chang, et al. 2000. Serologic and histopathologic study of Chlamydia pneumoniae infection in atherosclerosis: A possible pathogenetic mechanism of atherosclerosis induced by Chlamydia pneumoniae. *Yonsei Med J* 41 (3): 319–327.

Stanhope, K. L., and P. J. Havel. 2008. Fructose consumption: Potential mechanisms for its effects to increase visceral adiposity and induce dyslipidemia and insulin resistance. *Curr Opin Lipidol* 19 (1): 16–24.

Steinberg, D. 1997. Lewis A. Conner Memorial Lecture: Oxidative modification of LDL and atherogenesis. *Circulation* 95 (4): 1062–1071.

———. 2002. Atherogenesis in perspective: Hypercholesterolemia and inflammation as partners in crime. *Nat Med* 8 (11): 1211–1217.

———. 2004. Thematic review series: The pathogenesis of atherosclerosis. An interpretive history of the cholesterol controversy: Part I. *J Lipid Res* 45 (9): 1583–1593.

Steinberg, G., W. H. Slaton, Jr., D. R. Howton, and J. F. Mead. 1956. Metabolism of essential fatty acids. IV. Incorporation of linoleate into arachidonic acid. *J Biol Chem* 220 (1): 257–264.

Stenson, W. F., D. Cort, J. Rodgers, R. Burakoff, K. DeSchryver-Kecskemeti, T. L. Gramlich, and W. Beeken. 1992. Dietary supplementation with fish oil in ulcerative colitis. *Ann Intern Med* 116 (8): 609–614.

Stordy, B. J. 2000. Dark adaptation, motor skills, docosahexaenoic acid, and dyslexia. *Am J Clin Nutr* 71 (1 Suppl): 323S–326S.

Strange, P. G. 1992. *Brain biochemistry and brain disorders.* Oxford: Oxford University Press.

Strayer, D., M. Belcher, T. Dawson, B. Delaney, J. Fine, B. Flickenger, P. Friedman, et al. 2006. *Food fats and oils.* 9th ed. Washington, D.C.: Institute of Shortening and Edible Oils [accessed June 30, 2009]. http://www.iseo.org./foodfats.htm.

Strittmatter, W. J., and A. D. Roses. 1995. Apolipoprotein E and Alzheimer disease. *Proc Natl Acad Sci USA* 92 (11): 4725–4727.

Subbanagounder, G., N. Leitinger, D. C. Schwenke, J. W. Wong, H. Lee, C. Rizza, A. D. Watson, K. F. Faull, A. M. Fogelman, and J. A. Berliner. 2000. Determinants of bioactivity of oxidized phospholipids: Specific oxidized fatty acyl groups at the sn-2 position. *Arterioscler Thromb Vasc Biol* 20 (10): 2248–2254.

Tannenbaum, A. 1942. The genesis and growth of tumors. III. Effects of high fat diets. *Cancer Research* 2:468–475.

Taylor, K. E., C. J. Higgins, C. M. Calvin, J. A. Hall, T. Easton, A. M. McDaid, and A. J. Richardson. 2000. Dyslexia in adults is associated with clinical signs of fatty acid deficiency. *Prostaglandins Leukot Essent Fatty Acids* 63 (1–2): 75–78.

Temple, N. J. 1996. Dietary fats and coronary heart disease. *Biomed Pharmacother* 50 (6–7): 261–268.

Thiele, E. A. 2003. Assessing the efficacy of antiepileptic treatments: The ketogenic diet. *Epilepsia* 44 (Suppl 7): 26–29.

Thijssen, M. A., and Mensink. 2005. Fatty acids and atherosclerosis risk. In *Atherosclerosis: Diet and drugs,* ed. A. von Eckardstein, 170:165–194. Handbook of Experimental Pharmacology series. Berlin-Heidelberg: Springer-Verlag.

Thompson, C. M., and R. C. Grafström. 2008. Mechanistic considerations for formaldehyde-induced bronchoconstriction involving S-nitrosoglutathione reductase. *J Toxicol Environ Health A* 71 (3): 244–248.

Tokudome, S., M. Kojima, S. Suzuki, H. Ichikawa, Y. Ichikawa, M. Miyata, K. Maeda, et al. 2006. Marine n-3 fatty acids and colorectal cancer: Is there a real link? *Cancer Epidemiol Biomarkers Prev* 15 (2): 406–407.

Tsiplakou, E., and G. Zervas. 2008. The effect of dietary inclusion of olive tree leaves and grape marc on the content of conjugated linoleic acid and vaccenic acid in the milk of dairy sheep and goats. *J Dairy Res* 75 (3): 270–278.

Uauy, R., D. R. Hoffman, E. E. Birch, D. G. Birch, D. M. Jameson, and J. Tyson. 1994. Safety and efficacy of omega-3 fatty acids in the nutrition of very low birth weight infants: Soy oil and marine oil supplementation of formula. *J Pediatr* 124 (4): 612–620.

Ujike, H., and Y. Morita. 2004. New perspectives in the studies on endocannabinoid and cannabis: Cannabinoid receptors and schizophrenia. *J Pharmacol Sci* 96 (4): 376–381.

U.S. Census Bureau. 2000. *Statistical abstracts of the United States: The national data book.* Washington, DC: U.S. Census Bureau.

U.S. Census Bureau. 2005. Income, poverty, and health insurance coverage in the United States: 2004. Washington, DC: U.S. Government Printing Office [accessed June 30, 2009]. http://www.census.gov/prod/2005pubs/p60–229.pdf.

Vaddadi, K. S., C. J. Gilleard, R. H. Mindham, and R. Butler. 1986. A controlled trial of prostaglandin E1 precursor in chronic neuroleptic resistant schizophrenic patients. *Psychopharmacology* (Berl) 88 (3): 362–367.

Valenza, M., and E. Cattaneo. 2006. Cholesterol dysfunction in neurodegenerative diseases: Is Huntington's disease in the list? *Prog Neurobiol* 80 (4): 165–176.

van den Berg, E., J. M. Dekker, G. Nijpels, R. P. Kessels, L. J. Kappelle, E. H. de Haan, R. J. Heine, C. D. Stehouwer, and G. J. Biessels. 2008. Cognitive functioning in elderly persons with type 2 diabetes and metabolic syndrome: The Hoorn study. *Dement Geriatr Cogn Disord* 26 (3): 261–269.

Van Gaal, L., X. Pi-Sunyer, J. P. Despres, C. McCarthy, and A. Scheen. 2008. Efficacy and safety of rimonabant for improvement of multiple cardiometabolic risk factors in overweight/obese patients: Pooled 1-year data from the Rimonabant in Obesity (RIO) program. *Diabetes Care* 31 (Suppl 2): S229–S240.

Vane, J. R. 1971. Inhibition of prostaglandin synthesis as a mechanism of action for aspirin-like drugs. *Nat New Biol* 231 (25): 232–235.

Varadarajan, S., S. Yatin, M. Aksenova, and D. A. Butterfield. 2000. Review: Alzheimer's amyloid beta-peptide-associated free radical oxidative stress and neurotoxicity. *J Struct Biol* 130 (2–3): 184–208.

Veiraiah, A. 2005. Hyperglycemia, lipoprotein glycation, and vascular disease. *Angiology* 56 (4): 431–438.

Vetrivel, K. S., H. Cheng, W. Lin, T. Sakurai, T. Li, N. Nukina, P. C. Wong, H. Xu, and G. Thinakaran. 2004. Association of gamma-secretase with lipid rafts in post-Golgi and endosome membranes. *J Biol Chem* 279 (43): 44945–44954.

Visioli, F., C. Colombo, and C. Galli. 1998. Oxidation of individual fatty acids yields different profiles of oxidation markers. *Biochem Biophys Res Commun* 245 (2): 487–489.

von Euler, U. S. 1983. History and development of prostaglandins. *Gen Pharmacol* 14 (1): 3–6.

Vreugdenhil, M., C. Bruehl, R. A. Voskuyl, J. X. Kang, A. Leaf, and W. J. Wadman. 1996. Polyunsaturated fatty acids modulate sodium and calcium currents in CA1 neurons. *Proc Natl Acad Sci USA* 93 (22): 12559–12563.

Wang, Y. 2001. Cross-national comparison of childhood obesity: The epidemic and the relationship between obesity and socioeconomic status. *Int J Epidemiol* 30 (5): 1129–1136.

Warner, T. D., F. Giuliano, I. Vojnovic, A. Bukasa, J. A. Mitchell, and J. R. Vane. 1999. Nonsteroid drug selectivities for cyclo-oxygenase-1 rather than cyclo-oxygenase-2 are associated with human gastrointestinal toxicity: A full in vitro analysis. *Proc Natl Acad Sci USA* 96 (13): 7563–7568.

Watson, R. T., and J. E. Pessin. 2001. Subcellular compartmentalization and trafficking of the insulin-responsive glucose transporter, GLUT4. *Exp Cell Res* 271 (1): 75–83.

Wattenberg, L. W. 1983. Inhibition of neoplasia by minor dietary constituents. *Cancer Res* 43 (5 Suppl): 2448S–2453S.

Weber, W., and S. Newmark. 2007. Complementary and alternative medical therapies for attention-deficit/hyperactivity disorder and autism. *Pediatr Clin North Am* 54 (6): 983–1006.

Wentworth, P., Jr., J. Nieva, C. Takeuchi, R. Galve, A. D. Wentworth, R. B. Dilley, G. A. DeLaria, et al. 2003. Evidence for ozone formation in human atherosclerotic arteries. *Science* 302 (5647): 1053–1056.

Whitmer, R. A., E. P. Gunderson, E. Barrett-Connor, C. P. Quesenberry, Jr., and K. Yaffe. 2005. Obesity in middle age and future risk of dementia: A 27 year longitudinal population based study. *BMJ* 330 (7504): 1360.

WHO. 1984. WHO cooperative trial on primary prevention of ischaemic heart disease with clofibrate to lower serum cholesterol: Final mortality follow-up. Report of the Committee of Principal Investigators. *Lancet* 324 (8403): 600–604.

Willerson, J. T., and P. M. Ridker. 2004. Inflammation as a cardiovascular risk factor. *Circulation* 109 (21 Suppl 1): II 2–10.

WIN-NIDDK. 2007. Statistics related to overweight and obesity. Weight-Control Information Network, National Institute of Diabetes, Digestive and Kidney Diseases (NIDDK), National Institutes of Health, Washington, DC [accessed June 18, 2008]. http://win.niddk.nih.gov/statistics/index.htm.

Wolozin, B., W. Kellman, P. Ruosseau, G. G. Celesia, and G. Siegel. 2000. Decreased prevalence of Alzheimer disease associated with 3-hydroxy-3-methyglutaryl coenzyme A reductase inhibitors. *Arch Neurol* 57 (10): 1439–1443.

Wood, Philip A. 2006. *How fat works.* Cambridge, MA: Harvard University Press.

Woollett, L. A., D. K. Spady, and J. M. Dietschy. 1992. Saturated and unsaturated fatty acids independently regulate low density lipoprotein receptor activity and production rate. *J Lipid Res* 33 (1): 77–88.

WOSCOPS. 1998. Influence of pravastatin and plasma lipids on clinical events in the West of Scotland Coronary Prevention Study (WOSCOPS). *Circulation* 97 (15): 1440–1445.

Wright, J. M. 2002. The double-edged sword of COX-2 selective NSAIDs. *CMAJ* 167 (10): 1131–1137.

Wu, B. J., A. J. Hulbert, L. H. Storlien, and P. L. Else. 2004. Membrane lipids and sodium pumps of cattle and crocodiles: An experimental test of the membrane pacemaker theory of metabolism. *Am J Physiol Regul Integr Comp Physiol* 287 (3): R633–R641.

Yancy, W. S., Jr., M. K. Olsen, J. R. Guyton, R. P. Bakst, and E. C. Westman. 2004. A low-carbohydrate, ketogenic diet versus a low-fat diet to treat obesity and hyperlipidemia: A randomized, controlled trial. *Ann Intern Med* 140 (10): 769–777.

Yang, H., and C. Chen. 2008. Cyclooxygenase-2 in synaptic signaling. *Curr Pharm Des* 14 (14): 1443–1451.

Yang, H., J. Zhang, K. Andreasson, and C. Chen. 2008. COX-2 oxidative metabolism of endocannabinoids augments hippocampal synaptic plasticity. *Mol Cell Neurosci* 37 (4): 682–695.

Yavin, E., A. Brand, and P. Green. 2002. Docosahexaenoic acid abundance in the brain: A biodevice to combat oxidative stress. *Nutr Neurosci* 5 (3): 149–157.

Yermakova, A. V., J. Rollins, L. M. Callahan, J. Rogers, and M. K. O'Banion. 1999. Cyclooxygenase-1 in human Alzheimer and control brain: Quantitative analysis of expression by microglia and CA3 hippocampal neurons. *J Neuropathol Exp Neurol* 58 (11): 1135–1146.

Yoon, J. H., and S. J. Baek. 2005. Molecular targets of dietary polyphenols with anti-inflammatory properties. *Yonsei Med J* 46 (5): 585–596.

Young, L. R., and M. Nestle. 2002. The contribution of expanding portion sizes to the US obesity epidemic. *Am J Public Health* 92 (2): 246–249.

Young-Xu, Y., S. Jabbour, R. Goldberg, C. M. Blatt, T. Graboys, B. Bilchik, and S. Ravid. 2003. Usefulness of statin drugs in protecting against atrial fibrillation in patients with coronary artery disease. *Am J Cardiol* 92 (12): 1379–1383.

Yudkoff, M., Y. Daikhin, I. Nissim, O. Horyn, A. Lazarow, B. Luhovyy, and S. Wehrli. 2005. Response of brain amino acid metabolism to ketosis. *Neurochem Int* 47 (1–2): 119–128.

Zambell, K. L., N. L. Keim, M. D. Van Loan, B. Gale, P. Benito, D. S. Kelley, and G. J. Nelson. 2000. Conjugated linoleic acid supplementation in humans: Effects on body composition and energy expenditure. *Lipids* 35 (7): 777–782.

Zammit, V. A., I. J. Waterman, D. Topping, and G. McKay. 2001. Insulin stimulation of hepatic triacylglycerol secretion and the etiology of insulin resistance. *J Nutr* 131 (8): 2074–2077.

Zhang, W., T. Wang, Z. Pei, D. S. Miller, X. Wu, M. L. Block, B. Wilson, Y. Zhou, J. S. Hong, and J. Zhang. 2005. Aggregated alpha-synuclein activates microglia: A process leading to disease progression in Parkinson's disease. *FASEB J* 19 (6): 533–542.

Zhang, Y., R. Proenca, M. Maffei, M. Barone, L. Leopold, and J. M. Friedman. 1994. Positional cloning of the mouse obese gene and its human homologue. *Nature* 372 (6505): 425–432.

Zhao, W., P. S. Devamanoharan, and S. D. Varma. 2000. Fructose induced deactivation of antioxidant enzymes: Preventive effect of pyruvate. *Free Radic Res* 33 (1): 23–30.

Zhao, Z., D. J. Lange, A. Voustianiouk, D. MacGrogan, L. Ho, J. Suh, N. Humala, M. Thiyagarajan, J. Wang, and G. M. Pasinetti. 2006. A ketogenic diet as a potential novel therapeutic intervention in amyotrophic lateral sclerosis. *BMC Neurosci* 7:29.

Zock, P. L., and M. B. Katan. 1998. Linoleic acid intake and cancer risk: A review and meta-analysis. *Am J Clin Nutr* 68 (1): 142–153.

Zubay, Geoffrey. 1998. *Biochemistry*. 4th ed. Dubuque, IA: William C. Brown Publishers.

INDEX

ABOUT THE AUTHOR

GLEN LAWRENCE is a professor of chemistry and biochemistry at Long Island University in Brooklyn, New York. He received his Ph.D. degree in biochemistry at Utah State University and was a researcher at Tübingen University in Germany; University of California, Riverside; Mount Sinai School of Medicine in New York; and Columbia University, College of Physicians and Surgeons. He was a science adviser for the U.S. Food and Drug Administration, New York Regional Laboratories, and has carried out research at Long Island University. He has taught chemistry, biochemistry, neurochemistry, and advanced electives in the honors program at LIU. He has also spent sabbatical years teaching and doing research in East Africa, Southeast Asia, and Turkey.